D0712055

Sample Size Methodology

This is a volume in
STATISTICAL MODELING AND DECISION SCIENCE

Gerald J. Lieberman and Ingram Olkin, editors
Stanford University, Stanford, California

Sample Size Methodology

M. M. Desu
Department of Statistics
School of Medicine and Biomedical Sciences
State University of New York at Buffalo
Buffalo, New York

D. Raghavarao
Department of Statistics
School of Business and Management
Temple University
Philadelphia, Pennsylvania

ACADEMIC PRESS, INC.
Harcourt Brace Jovanovich, Publishers

Boston San Diego New York
London Sydney Tokyo Toronto

ARLIS
Alaska Resources Library & Information Services
Library Building, Suite 111
3211 Providence Drive
Anchorage, AK 99508-4614

ACADEMIC PRESS, INC.
1250 Sixth Avenue, San Diego, CA 92101

United Kingdom Edition published by
ACADEMIC PRESS LIMITED
24–28 Oval Road, London NW1 7DX

Library of Congress Cataloging-in-Publication Data

Desu, M. M.
    Sample size methodology/M. M. Desu, D. Raghavarao.
        p.   cm.—(Statistical modeling and decision science)
    Includes bibliographical references.
    ISBN 0-12-212165-1 (alk. paper)
    1. Sampling (Statistics) 2. Distribution (Probability theory)
I. Raghavarao, Damaraju.   II. Title.   III. Series.
QA27.6.D47   1990
519.5′2—dc20   90-400
                                                                       CIP

Printed in the United States of America

90 91 92 93   9 8 7 6 5 4 3 2 1

To my wife, Aruna Desu
—M. M. Desu

In loving memory of my wife, Damaraju V. Rathnam
—D. Raghavarao

Contents

Preface xi

Chapter 1   One-Sample Problems 1

1.1 Introduction   1
1.2 Sampling from a Normal Distribution   2
    1.2.1 Estimation of μ When σ^2 Is Known   3
    1.2.2 Estimation of μ When σ^2 Is Unknown   4
    1.2.3 Estimation of σ^2 5
    1.2.4 Tests of Hypothesis about μ When σ^2 Is Known   7
    1.2.5 Tests of Hypothesis about μ When σ^2 Is Unknown   9
    1.2.6 Tests of Hypothesis about σ^2 11
1.3 Sampling from a Bernoulli Distribution   12
    1.3.1 Estimation of θ 13
    1.3.2 Tests of Hypothesis about θ 15
1.4 Sampling from an Exponential Distribution   16
    1.4.1 Estimation of γ When θ Is Known   17
    1.4.2 Estimation of γ When θ Is Unknown   17
    1.4.3 Estimation of θ 18
    1.4.4 Tests of Hypothesis about θ 19
1.5 Bayesian Approach to Estimation   20

Chapter 2   Two-Sample Problems 23

2.1 Introduction   23
2.2 Sampling from Two Normal Distributions   23
    2.2.1 Estimation of $\mu_1 - \mu_2$ When σ_1^2, σ_2^2 Are Known   24

2.2.2 Estimation of $\mu_1 - \mu_2$ When $\sigma_1^2 = \sigma_2^2$ and the Common Variance Is Unknown   25
2.2.3 Estimation of $\mu_1 - \mu_2$ When $\sigma_1^2 \neq \sigma_2^2$ and the Variances Are Unknown   26
2.2.4 Estimation of Variance Ratio, σ_1^2/σ_2^2 27
2.2.5 Tests of Hypothesis about μ_1 and μ_2 When σ_1^2 and σ_2^2 Are Known   30
2.2.6 Tests of Hypothesis about μ_1 and μ_2 When $\sigma_1^2 = \sigma_2^2$ and the Common Variance Is Unknown   31
2.2.7 Tests of Hypothesis about μ_1 and μ_2 When $\sigma_1^2 \neq \sigma_2^2$ and the Variances Are Unknown   33
2.2.8 Tests of Hypothesis about σ_1^2 and σ_2^2 34
2.3 Sampling from Two Bernoulli Distributions   36
2.4 Sampling from Two Exponential Distributions   37
    2.4.1 Estimation of θ_1/θ_2 37
    2.4.2 Tests of Hypothesis about θ_1/θ_2 38

Chapter 3   _k_-Sample Problems   41

3.1 Introduction   41
3.2 Estimation of Difference of Means   42
    3.2.1 Estimation without Distributional Assumptions   42
    3.2.2 Estimation under Normality Assumption   44
3.3 Sampling from Normal Distributions   46
3.4 Sampling from Bernoulli Distributions   49
3.5 Sampling from One-Parameter Exponential Distributions   51

Chapter 4   Power and Sample Size   55

4.1 Introduction   55
4.2 Noncentral t-Distribution   56
    4.2.1 Power Function of the One-Sample t-Test   56
    4.2.2 Power Function of the Two-Sample-t-Test   58
4.3 Noncentral χ^2-Distribution   59
4.4 Noncentral F-Distribution   60

Chapter 5   Samples from Finite Populations   63

5.1 Introduction   63
5.2 Simple Random Sampling   63
    5.2.1 Acceptance Sampling   66

5.3 Stratified Random Sampling 67
5.4 Double Sampling for Stratification 69
5.5 Two-Stage Cluster Sampling 70

Chapter 6   Ranking and Selection 73

6.1 Introduction 73
6.2 Selecting the Best Normal Distribution 74
    6.2.1 Selecting the Normal Distribution with the Largest Mean When the Common Variance σ^2 Is Known 74
    6.2.2 Selecting the Normal Distribution with the Largest Mean When the Common Variance σ^2 Is Unknown 76
    6.2.3 Selecting the Normal Distribution with the Smallest Variance 76
6.3 Selecting the Best Bernoulli Distribution 78
6.4 Selecting the Best Exponential Distribution 80

Chapter 7   Biomedical Experiments and Clinical Trials 83

7.1 Introduction 83
7.2 Experiments with Qualitative Response Variable 83
    7.2.1 McNemar's Test 84
    7.2.2 Testing Equality of r Treatment Effects with c Categorical Responses 88
    7.2.3 Long-Term Trials 91
    7.2.4 Case Control Studies 91
    7.2.5 Two-Stage Designs 94
7.3 Experiments with Quantitative Response Variable 94
    7.3.1 Testing for Bioequivalence 95
    7.3.2 Two Period, Two Treatment, Crossover Study 97
    7.3.3 Clinical Trials for Comparison of Survival Distributions 99

Appendix: Guide to Selected Sample Size Tables 103

Problems and Supplements 109

References 125

Index 133

Preface

Sample size determination is one of the important problems in designing an experiment or a survey. The general results of this problem are scattered in various textbooks and journal articles. There seems to have been no comprehensive monograph on this topic, except for some specialized coverage in books, such as "Sample Size Determination," by Mace, "Statistical Power Analysis," by Cohen, and "How to Choose the Proper Sample Size," by Brush.

In this monograph, the authors present the currently available methodology for the purpose of estimation and tests of hypotheses. This discussion includes the two cases, namely, random sampling from standard probability distributions and from finite populations. Sample size determination for estimating parameters in a Bayesian setting by considering the posterior distribution of the parameter and specifying the necessary requirements is also briefly discussed. Further, the determination of the sample size is also considered for ranking and selection problems as well as for design of clinical trials.

In relation to the estimation problems, the sample size is determined by controlling the absolute error or relative error with a high degree of probability based on the feasibility of the procedure when the sampling is from known probability distributions. When the samples are from a finite population, the variance of the estimator is controlled, and the sample size is obtained. It may be noted that controlling the variance of an estimator can be related to controlling the absolute error of the estimator. The sample size in the context of testing a hypothesis is determined by controlling the power of an α-level test for certain alternatives.

A two-semester mathematical statistics course provides an adequate background for understanding the material presented in this monograph. As such, this monograph can be used as a text for a senior-level course on this topic. This material has been successfully used as a text in a graduate course given by one of the authors.

This monograph is not a ready reckoner for someone to determine the sample size for a specific problem and, as such, no tables or charts are provided. However, it contains appropriate techniques for attacking the general question of sample size determination in problems of estimation, tests of hypotheses, selection, and clinical trial design. It will aid the reader in formulating an appropriate problem of sample size and in obtaining the solution. This may also serve as a reference book for consulting statisticians.

An annotated list of tables is given in the appendix, which is a ready reference for the reader in determining the sample sizes for the problems discussed in the text. A collection of problems is given at the end of the book, and this set will serve as a useful supplement to the material covered in the text.

The authors wish to thank Ms. Shelah Burgess for her efficient typing of the original manuscript.

December 1989   M. M. Desu
                                                                   D. Raghavarao

Chapter 1   One-Sample Problems

1.1   Introduction

Statisticians draw inferences about population parameters based on samples of appropriate size. The number of units used in the sample is called the sample size.

Consider a gubernatorial election in which Candidate A has actually received 45% of the electoral votes. If an exit poll of 100 voters is taken, then there is a 13.35% chance of his getting more than half of the votes. If a sample of 2000 voters is taken, however, then this chance is nearly zero. By taking a small sample of 100 voters, the candidate gets a false hope of winning the election, while the *true* picture emerges with a large sample of voters such as 2000. Interviewing 2000 voters is expensive and time consuming. There is a minimum required sample size that enables one to estimate the proportion of his preferred votes to the desired level of accuracy.

Again, take the case of a manufacturer of Model X cars with given equipment options who claims that those cars give an average gas mileage of at least 25 mpg. Assume that the standard deviation of gas mileage delivered by such cars is 1 mpg. A consumer protection agency wants to disprove the manufacturer's claim using a test of .05 level of significance. Using only four test cars, the agency has a 26% chance of rejecting the manufacturer's claim, when the actual gas mileage delivered is 24.5. Thus, cars giving $\frac{1}{2}$ mpg less than the claim have a 74% chance of meeting the manufacturer's claim. If 100 cars are tested, this chance of accepting the manufacturer's claim becomes nearly 0%. Then once again there is an appropriate sample size that enables the experimenter even to retain a hypothesis with a high probability when it is false.

In this chapter, necessary tools will be developed to provide answers for questions on appropriate sample sizes in different settings, when inferences are drawn about the parameter (or parameters) of a single population using only one random sample. At present, it is assumed that the random samples are drawn from probability distributions, and problems involving finite populations will be considered in Chapter 5.

1.2   Sampling from a Normal Distribution

Let X be a random variable following a normal distribution with mean μ and variance σ^2 having the density function

$$f(x; \mu, \sigma^2) = \frac{1}{\sqrt{2\pi\sigma^2}} \exp\left(-\frac{(x-\mu)^2}{2\sigma^2}\right), \qquad -\infty < x < \infty. \quad (1.2.1)$$

This will be written as $X \sim N(\mu, \sigma^2)$. Let $X_1, X_2, \ldots, X_n$ be a random sample on X from the normal distribution. Further, let

$$\bar{X} = \sum_{i=1}^{n} \frac{X_i}{n}, \qquad s^2 = \sum_{i=1}^{n} \frac{(X_i - \bar{X})^2}{(n-1)}. \quad (1.2.2)$$

The sample mean $\bar{X}$ is known to have a normal distribution with mean μ and variance σ^2/n and is independently distributed from s^2. The quantity $Y = (n-1)s^2/\sigma^2$ is known to have a chi-square

(χ^2) distribution with $\nu = n - 1$ degrees of freedom, denoted by $\chi^2(\nu)$, having the density function.

$$f(y; \nu) = \frac{1}{\Gamma\left(\dfrac{\nu}{2}\right) 2^{\nu/2}} y^{\nu/2-1} e^{-y/2}, \qquad 0 < y < \infty. \qquad (1.2.3)$$

Further, the quantity

$$T = \frac{\bar{X} - \mu}{\dfrac{s}{\sqrt{n}}}, \qquad (1.2.4)$$

has Student's t-distribution with $\nu = n-1$ degrees of freedom, denoted by $T(\nu)$, having density function

$$f(t; \nu) = \frac{\Gamma\left(\dfrac{\nu+1}{2}\right)}{\sqrt{\pi\nu}\,\Gamma\left(\dfrac{\nu}{2}\right)} \frac{1}{\left(1 + \dfrac{t^2}{\nu}\right)^{(\nu+1)/2}}. \qquad (1.2.5)$$

The required sample size for estimating and testing hypothesis about the parameters μ or σ^2 are considered in the following sections.

1.2.1 Estimation of μ When σ^2 Is Known

It is known that the best unbiased estimator of μ is $\bar{X}$. The absolute error in estimating μ by $\bar{X}$ is $|\bar{X} - \mu|$. It is desirable to control this absolute error at a certain level. Mathematically, this requirement can be stated as

$$P(|\bar{X} - \mu| \leq d) \geq 1 - \alpha, \qquad (1.2.6)$$

where d and α are prechosen positive constants. It is possible to choose n suitably to meet the condition given in (1.2.6), and the result is contained in the following:

Theorem 1.2.1 *The sample size n, required to meet the inequality* (1.2.6) *is* $[n^*] + 1$, *where*

$$n^* = \left(z_{\alpha/2}\frac{\sigma}{d}\right)^2, \qquad (1.2.7)$$

[·] *is the greatest integer function, and* $z_{\alpha/2}$ *is the upper* $100(\alpha/2)$ *percentile point of the standard normal distribution.*

Example 1.2.1.1   Consider the problem of estimating the gas mileage of a given brand of cars with specific options, such that a maximum absolute error of $\frac{1}{2}$ mpg is allowed with a probability of at least 0.95. Furthermore, it is known that the standard deviation of the gas mileage distribution is 1 mpg. This requirement can be expressed as

$$P(|\bar{X} - \mu| \leqslant 0.5) \geqslant 0.95,$$

and the required number of cars to be tested, from Theorem 1.2.1, is 16, because

$$n^* = \left\{\frac{(1.96)(1)}{0.5}\right\}^2 = 15.4.$$

1.2.2   Estimation of μ When σ^2 Is Unknown

In this case, if the accuracy d given in (1.2.6) is expressed in units of σ, that is, $k\sigma$, where k is specified, (1.2.7) is applicable and yields

$$n^* = \left(\frac{z_{\sigma/2}}{k}\right)^2. \qquad (1.2.8)$$

If such a specification of $d = k\sigma$ is not feasible, then one may have to consider a modification of Stein's two-stage procedure to estimate μ as described next:

(i) Take an initial sample of size $n_1(>2)$. Let s_1^2 be the variance of this sample.

(ii) Let $n = \max \left\{ n_1, \left[\frac{s_1^2 t_{\alpha/2}^2(n_1 - 1)}{d^2}\right] + 1 \right\}$, where $t_p(\nu)$ is the upper $100p$ percentile of the t-distribution with ν degrees of freedom.

(iii) Take $n - n_1$ additional observations, if necessary.

(iv) Let $\bar{X}$ be the mean of all n observations.

It is then known that the estimator $\bar{X}$ satisfies the requirement given in (1.2.6).

The question to be resolved in these two-stage procedures is the choice of n_1, the initial sample size. One can evaluate $E(n)$ for different n_1 and σ^2 values and make a judicious choice of n_1. For a detailed discussion of this aspect of these procedures, see Moshman (1958) and Seelbinder (1953).

Consider the Example 1.2.1.1 and suppose that the standard deviation, σ, of the gas mileage distribution is unknown. The experimenter may initially take $n_1 = 5$, say, cars and let the gas mileages delivered by the five cars be 24.5, 26, 25, 24.9, and 24.6 mpg. Here, $s_1^2 = 0.355$. Now

$$n = \max\left\{5, \left[\frac{(2.776)^2(0.355)}{(0.5)^2}\right] + 1\right\} = 11.$$

Thus, the experimenter has to take $11 - 5 = 6$ additional cars and collect the data on them to meet the specifications for estimating the mean.

1.2.3 *Estimation of σ^2*

An unbiased estimator of σ^2 is known to be

$$\hat{\sigma}^2 = \sum_{i=1}^{n} \frac{(X_i - \mu)^2}{n}, \qquad \text{if } \mu \text{ is known;}$$

$$= s^2, \qquad \text{if } \mu \text{ is unknown.} \tag{1.2.9}$$

Let $f = n$ or $n - 1$ depending on whether μ is known or unknown. The quantity $f\hat{\sigma}^2/\sigma^2$ is distributed as $\chi^2(f)$.

It is not possible to determine the sample size by controlling the absolute error as in the case of estimating the mean. It is easy to see that the $P(|\hat{\sigma}^2 - \sigma^2| \leq d)$ cannot be evaluated without prior information on σ^2. If d is taken as $r\sigma^2$, however, then $P(|\hat{\sigma}^2 - \sigma^2| \leq r\sigma^2) = P(|(\hat{\sigma}^2 - \sigma^2)/\sigma^2| \leq r)$, and this probability can be evaluated. This modified specification controls the relative error $|(\hat{\sigma}^2 - \sigma^2)/\sigma^2|$. Thus, n will be found to satisfy the requirement

$$P\left(\left|\frac{\hat{\sigma}^2 - \sigma^2}{\sigma^2}\right| \leq r\right) \geq 1 - \alpha. \tag{1.2.10}$$

This criterion is equivalent to

$$P\left[(1-r)f \le \frac{f\hat{\sigma}^2}{\sigma^2} \le (1+r)f\right] \ge 1 - \alpha. \qquad (1.2.11)$$

Let $G_f(\cdot)$ be the distribution function of a χ^2 variable with f degrees of freedom. Equation (1.2.11) can be written as

$$G_f((1+r)f) - G_f((1-r)f) \ge 1 - \alpha. \qquad (1.2.12)$$

Theorem 1.2.2   *The required sample size n that satisfies the spec-ification (1.2.10) is f or f + 1 according to whether μ is known or unknown, where f is the smallest positive integer that satisfies (1.2.12).*

Since the inequality (1.2.12) cannot be solved explicitly for f, a rough approximation to obtain f will now be given. From the Central Limit Theorem, the distribution of $(\chi^2(f) - f)/\sqrt{2f}$ is asymptotically $N(0, 1)$. Using this fact, from Equation (1.2.12) one gets

$$\Phi\left(r\sqrt{\frac{f}{2}}\right) - \Phi\left(-r\sqrt{\frac{f}{2}}\right) \ge 1 - \alpha, \qquad (1.2.13)$$

where $\Phi(\cdot)$ is the distribution function of the standard normal variable. This can be easily solved for f, resulting in the following theorem:

Theorem 1.2.3   *An approximation to the sample size n satisfying the criterion (1.2.10) is $[f^*] + 1$ or $[f^*] + 2$ accordingly as μ is known or unknown, where*

$$f^* = \frac{2z_{\alpha/2}^2}{r^2}. \qquad (1.2.14)$$

Example 1.2.3.1   An engineer is interested in estimating the variance in resistance values of a certain resistor type controlling the relative error at 20% with a probability of at least 95%. Here, $r = .20$ and $1 - \alpha = .95$. Thus, (1.2.14) gives

$$f^* = \frac{2(1.96)^2}{(0.2)^2} = 192.08$$

The required sample size is 193 or 194 depending on whether μ is known or unknown to the engineer.

1.2.4   Tests of Hypothesis about μ When σ^2 Is Known

Consider the problem of testing a null hypothesis $H_0: \mu = \mu_0$ against the one-sided alternative $H_A: \mu > \mu_0$ with α level of significance. The critical region is known to be

$$\bar{X} > \mu_0 + z_\alpha\left(\frac{\sigma}{\sqrt{n}}\right). \tag{1.2.15}$$

The sample size will now be determined so that the test has a specified power of $1 - \beta$ at the alternative $\mu = \mu_1(>\mu_0)$. In other words, n is chosen such that

$$P(\bar{X} > \mu_0 + z_\alpha(\sigma/\sqrt{n})|\mu = \mu_1) = 1 - \beta. \tag{1.2.16}$$

This equation yields $n = [n^*] + 1$, where

$$n^* = \left\{\frac{\sigma}{(\mu_1 - \mu_0)}(z_\alpha + z_\beta)\right\}^2. \tag{1.2.17}$$

This result is given in the following:

Theorem 1.2.4   *The sample size n required to give a power $1 - \beta$ at the alternative $\mu = \mu_1$ ($>\mu_0$) of an α-level one-sided test of H_0: $\mu = \mu_0$ is $[n^*] + 1$, where n^* is given by (1.2.17).*

Example 1.2.4.1   The average milk yield of cows with a standard feed is 1450 lbs over a 10-week period with a standard deviation of 70 lbs. A new feed is to be tested, and the new feed is expected to increase the yield. The investigator uses a .05 level one-sided test and is interested to have a power of .90 when the real mean yield is 1500 lbs. The required sample size is 17 because

$$n^* = \left\{\frac{70(1.645 + 1.282)}{50}\right\}^2 = 16.79.$$

The sample size given in Theorem 1.2.4 is also useful in the context of testing $H_0:\mu = \mu_0$ against the one-sided alternative $H_A: \mu < \mu_0$.

Now, consider the problem of testing the null hypothesis $H_0: \mu = \mu_0$ against the two-sided alternative $H_A: \mu \neq \mu_0$ using an α-level test. The critical region is

$$|\bar{X} - \mu_0| > z_{\alpha/2}\left(\frac{\sigma}{\sqrt{n}}\right). \tag{1.2.18}$$

The sample size n will be found so that the test has a specified power $1 - \beta$ at the alternative $\mu = \mu_1 \ (\neq \mu_0)$. The sample size n is thus chosen so that

$$P\left(|\bar{X} - \mu_0| > z_{\alpha/2}\left(\frac{\sigma}{\sqrt{n}}\right)\Big|\mu = \mu_1\right) = 1 - \beta. \tag{1.2.19}$$

This can be rewritten as

$$\Phi\left(-z_{\alpha/2} - \frac{(\mu_1 - \mu_0)}{\sigma/\sqrt{n}}\right) + 1 - \Phi\left(z_{\alpha/2} - \frac{(\mu_1 - \mu_0)}{\sigma/\sqrt{n}}\right) = 1 - \beta, \tag{1.2.20}$$

If $\mu_1 > \mu_0$, then the first $\Phi(\cdot)$ term on the left-hand side of (1.2.20) is close to 0, and an approximate n, a solution of (1.2.20) satisfies

$$z_{\alpha/2} - \frac{(\mu_1 - \mu_0)}{\sigma/\sqrt{n}} = -z_\beta. \tag{1.2.21}$$

On the other hand, if $\mu_1 < \mu_0$, then the second $\Phi(\cdot)$ term on the left-hand side of (1.2.20) is close to 1, and an approximate n, a solution of (1.2.20) satisfies

$$-z_{\alpha/2} - \frac{(\mu_1 - \mu_0)}{\sigma/\sqrt{n}} = z_\beta. \tag{1.2.22}$$

Equations (1.2.21) and (1.2.22) can be combined into one equation

$$z_{\alpha/2} + z_\beta = \frac{|\mu_1 - \mu_0|}{\sigma/\sqrt{n}}. \tag{1.2.23}$$

Hence, n is $[n^*] + 1$, where

$$n^* = \left\{ \frac{(z_{\alpha/2} + z_\beta)\sigma}{|\mu_1 - \mu_0|} \right\}^2.$$ (1.2.24)

The following theorem is thus established.

Theorem 1.2.5 *The sample size n required to give a power of $1 - \beta$ at the alternative $\mu = \mu_1$ $(\neq \mu_0)$ to an α-level two-sided test of H_0: $\mu = \mu_0$ is $[n^*] + 1$, where n^* is given by (1.2.24).*

1.2.5   Tests of Hypothesis about μ When σ^2 Is Unknown

In this case, the critical region of a one-sided α-level test for testing H_0: $\mu = \mu_0$ against the alternative H_A: $\mu > \mu_0$ is

$$\bar{X} > \mu_0 + t_\alpha(n - 1)\frac{s}{\sqrt{n}}.$$ (1.2.25)

By requiring a power of $1 - \beta$ at $\mu = \mu_1$ $(>\mu_0)$, one gets

$$P(\bar{X} > \mu_0 + t_\alpha(n - 1)\frac{s}{\sqrt{n}}|\mu = \mu_1) = 1 - \beta.$$ (1.2.26)

Approximating s by σ, (1.2.26) can be simplified to

$$P\left(T(n - 1) > t_\alpha(n - 1) - \sqrt{n}\,\frac{(\mu_1 - \mu_0)}{\sigma} \right) = 1 - \beta,$$ (1.2.27)

and thus

$$t_\alpha(n - 1) - \sqrt{n}\,\frac{(\mu_1 - \mu_0)}{\sigma} = -t_\beta(n - 1).$$ (1.2.28)

If σ_u is a known upper bound for the unknown σ, (1.2.28) can be solved for n iteratively. It is the smallest positive integer satisfying

$$n = \left\{ \frac{\sigma_u(t_\alpha(n - 1) + t_\beta(n - 1))}{(\mu_1 - \mu_0)} \right\}^2.$$ (1.2.29)

It is somewhat unrealistic to assume that an upper bound σ_u will be known for the unknown σ. Furthermore, if σ_u is set at the extreme, n will be very large, thus defeating the purpose of sample size consideration. To make things more complicated, in the process of deriving (1.2.29), s was approximated by σ, which may be valid only for large n. All of these considerations tend to undermine the use of (1.2.29) as an appropriate answer to the sample size question in this case.

Alternatively, satisfactory results can be obtained using a two-stage t-test as described next:

(i) Take an initial sample of size $n_1(>2)$. Let s_1^2 be the variance of this sample.

(ii) Let

$$c = \left\{ \frac{\mu_1 - \mu_0}{t_\alpha(n_1 - 1) + t_\beta(n_1 - 1)} \right\}^2. \qquad (1.2.30)$$

(iii) Let $n = \max\left\{ n_1, \left[\dfrac{s_1^2}{c} \right] + 1 \right\}$.

(iv) Take $n - n_1$ additional observations, if necessary. Let $\bar{X}$ be the sample mean of the total sample of size n.

(v) The critical region of a one-sided α-level test for testing H_0: $\mu = \mu_0$ against the alternative H_A: $\mu > \mu_0$ giving a power of at least $1 - \beta$ at $\mu = \mu_1$ is

$$\bar{X} > \mu_0 + t_\alpha(n_1 - 1)\left(\frac{s_1}{\sqrt{n}} \right). \qquad (1.2.31)$$

The mathematical details in the development of this procedure are discussed in Problem 7 of the Problems and Supplements part of this monograph. A similar approach can be taken for testing H_0: $\mu = \mu_0$ against H_A: $\mu < \mu_0$. For testing H_0: $\mu = \mu_0$ against the two-sided alternative H_A: $\mu \neq \mu_0$ using a two-stage procedure, replace (1.2.30) by

$$c = \left\{ \frac{|\mu_1 - \mu_0|}{t_{\alpha/2}(n_1 - 1) + t_\beta(n_1 - 1)} \right\}^2, \qquad (1.2.32)$$

and replace (1.2.31) by

$$|\bar{X} - \mu_0| > t_{\alpha/2}(n_1 - 1)\left(\frac{s_1}{\sqrt{n}} \right). \qquad (1.2.33)$$

Consider Example 1.2.4.1 and let σ be unknown. Initially, the experimenter may select a sample of $n_1 = 5$ cows and let the milk yields for the five cows for the 10-week period be 1490, 1560, 1460, 1480, and 1510. Here, $s_1^2 = 1450$. From Equation (1.2.30), using $\alpha = .05$, $\beta = .10$, and $\mu_1 - \mu_0 = 50$, one gets $c = 186.12$.

$$n = \max\left\{5, \left[\frac{1450}{186.12}\right] + 1\right\} = 8.$$

Thus, the data on three more cows are needed to meet the specification.

1.2.6 *Tests of Hypothesis about σ^2*

Using the notation of Section 1.2.3, the commonly tested hypothesis about σ^2 is H_0: $\sigma^2 = \sigma_0^2$ against the one-sided alternative H_A: $\sigma^2 > \sigma_0^2$. The critical region of an α-level test is

$$\hat{\sigma}^2 > \frac{\sigma_0^2 \chi_{1-\alpha}^2(f)}{f}, \tag{1.2.34}$$

where $\chi_{1-\alpha}^2(f)$ is the $100(1 - \alpha)$ percentile point of $\chi^2(f)$. For having a power of $1 - \beta$ when $\sigma^2 = \sigma_1^2$ ($>\sigma_0^2$), the sample size n must be chosen so that

$$P\left(\hat{\sigma}^2 > \frac{\sigma_0^2 \chi_{1-\alpha}^2(f)}{f} \;\middle|\; \sigma^2 = \sigma_1^2\right) = 1 - \beta. \tag{1.2.35}$$

This leads to the relation

$$\sigma_0^2 \chi_{1-\alpha}^2(f) = \sigma_1^2 \chi_{\beta}^2(f), \tag{1.2.36}$$

which needs to be solved for f. Thus,

Theorem 1.2.6 *The sample size n required to give a power of $1 - \beta$ at the alternative $\sigma^2 = \sigma_1^2$ ($>\sigma_0^2$) to an α-level one-sided test of H_0: $\sigma^2 = \sigma_0^2$ is f or $f + 1$ depending on whether μ is known or unknown, where f is the smallest positive integer satisfying* (1.2.36).

The χ^2 percentiles can be approximated by normal percentiles as

$$\chi_p^2(f) = (\Phi^{-1}(p) + \sqrt{2f - 1})^2 \cdot \tfrac{1}{2}.$$

Using this approximation in (1.2.36) and solving for f gives the solution

$$f^* = \frac{1}{2} \left(\frac{\lambda z_\alpha + z_\beta}{\lambda - 1} \right)^2 + \frac{1}{2}, \tag{1.2.37}$$

where $\lambda^2 = \sigma_0^2 / \sigma_1^2$. Thus,

Theorem 1.2.7 *An approximate sample size n required to give a power of $1 - \beta$ at the alternative $\sigma^2 = \sigma_1^2$ ($> \sigma_0^2$) to an α-level one-sided test of H_0: $\sigma^2 = \sigma_0^2$ is $[f^*] + 1$ or $[f^*] + 2$ depending on whether μ is known or unknown.*

Example 1.2.7.1 An engineer wants to test the variability in resistance values for certain resistors. He wants to perform a 0.05 level one-sided test for H_0: $\sigma^2 = 2,000$, and he wants a power of 0.90 when $\sigma^2 = 3,125$. Here $\lambda = 0.80$, and

$$f^* = \frac{1}{2} \left\{ \frac{(0.80)(1.645) + 1.282}{0.80 - 1} \right\}^2 + \frac{1}{2}$$

$$= 84.87.$$

The required sample size for this problem is 85 or 86 depending on whether μ is known or unknown.

Determination of sample size in the context of constructing tolerance intervals can be related to the same problem in the context of testing a hypothesis. Such a discussion appears in Guenther (1977, Chapter 4). For additional discussion and tables, one may refer to Odeh, *et al.* (1987, 1989).

1.3   Sampling from a Bernoulli Distribution

Let X be a discrete random variable with probability function

$$f(x; \theta) = \theta^x (1 - \theta)^{1-x}, x = 0, 1. \tag{1.3.1}$$

It is known that the mean and variance of this distribution are, respectively, θ and $\theta(1 - \theta)$. Let $X_1, X_2, \ldots, X_n$ be a random sample on X. Then $Y = \Sigma_{i=1}^{n} X_i$ is known to have a binomial distribution with parameters n and θ, whose probability function is

$$f(y; n, \theta) = \binom{n}{y} \theta^y (1 - \theta)^{n-y}, \qquad y = 0, 1, \ldots, n. \quad (1.3.2)$$

The mean and variance of Y are $n\theta$ and $n\theta(1 - \theta)$, respectively. Thus, Y/n is an unbiased estimator of θ.

The required sample size for estimating and testing hypotheses on θ will be considered in the following sections.

1.3.1   Estimation of θ

To determine the sample size n for estimating θ by $\hat{\theta} = Y/n$ controlling the absolute error with a high probability, one needs to choose n to satisfy

$$P(|Y/n - \theta| \leq d) \geq 1 - \alpha, \qquad (1.3.3)$$

for given positive constants d and α. The left-hand side of (1.3.3) is

$$P([n(\theta - d)] + 1 \leq Y \leq [n(\theta + d)]) \equiv P(y_1 \leq Y \leq y_2)$$

$$= \sum_{y=y_1}^{y_2} \binom{n}{y} \theta^y (1 - \theta)^{n-y}.$$

$$(1.3.4)$$

Since this probability depends on the unknown parameter θ, one may note that it is least when $\theta = 0.5$. Thus, a generous sample size n can be determined by setting $\theta = 0.5$ in (1.3.4) and making it not less than $1 - \alpha$.

To obtain a quick solution to (1.3.3), the probability of (1.3.4) will be approximated. From Central Limit Theorem, the distribution of $(Y/n - \theta)/\sqrt{\theta(1 - \theta)/n}$ is asymptotically $N(0, 1)$.

Using this fact, from Equation (1.3.3) one gets

$$\Phi\left(\frac{d}{\sqrt{\frac{\theta(1-\theta)}{n}}}\right) - \Phi\left(\frac{-d}{\sqrt{\frac{\theta(1-\theta)}{n}}}\right) \geq 1 - \alpha,$$

i.e.,

$$\Phi\left(\frac{d}{\sqrt{\frac{\theta(1-\theta)}{n}}}\right) \geq 1 - \frac{\alpha}{2}. \qquad (1.3.5)$$

Thus, n must satisfy

$$n \geq \frac{\theta(1-\theta)z_{\alpha/2}^2}{d^2}. \qquad (1.3.6)$$

Since θ is unknown and $\theta(1-\theta)$ is maximum, when $\theta = 0.5$, one may use a generous estimate of n as $[n^*] + 1$, where

$$n^* = \frac{z_{\alpha/2}^2}{4d^2}. \qquad (1.3.7)$$

Thus,

Theorem 1.3.1   *An approximate sample size n for estimating θ, the probability of success of a Bernoulli distribution, by $\hat{\theta} = Y/n$ satisfying the requirement (1.3.3) is $[n^*] + 1$, where n^* is given by (1.3.7).*

Example 1.3.1.1   A pharmacologist wants to estimate the response rate of a drug to within 0.03 with a probability of at least 95%. Here, $d = 0.03$, and $1 - \alpha = 0.95$, so that

$$n^* = \frac{(1.96)^2}{4(0.03)^2}$$

$$= 1067.11.$$

The required sample size in this problem is $n = 1068$.

1.3.2 Tests of Hypothesis about θ

Consider an α-level test of H_0: $\theta = \theta_0$ against the one-sided alternative H_A: $\theta > \theta_0$. The critical region is

$$Y > c, \tag{1.3.8}$$

where c satisfies

$$P(Y > c \mid \theta = \theta_0) \leq \alpha, \tag{1.3.9}$$

or equivalently,

$$I_{\theta_0}(c + 1, n - c) \leq \alpha, \tag{1.3.10}$$

$I_x(a, b)$ being the usual incomplete beta function

$$I_x(a, b) = \frac{\int_0^x u^{a-1}(1 - u)^{b-1}\, du}{\int_0^1 u^{a-1}(1 - u)^{b-1}\, du}. \tag{1.3.11}$$

By requiring a power of at least $1 - \beta$ at $\theta = \theta_1$ $(>\theta_0)$, one gets

$$P(Y > c \mid \theta = \theta_1) \geq 1 - \beta, \tag{1.3.12}$$

from which the sample size n is determined. Thus, n and c must satisfy the following two equations:

$$I_{\theta_0}(c + 1, n - c) \leq \alpha, \qquad I_{\theta_1}(c + 1, n - c) \geq 1 - \beta. \tag{1.3.13}$$

These equations have to be solved iteratively.

An approximate solution to n can be obtained by making the arcsin transformation on Y/n and expressing the probabilities in Equations (1.3.9) and (1.3.12) in terms of standard normal distribution function. It is known that

$$Z = 2\sqrt{n}\,(\arcsin \sqrt{\hat{\theta}} - \arcsin \sqrt{\theta})$$

is asymptotically distributed as $N(0, 1)$. It may be noted that the arcsin $(\cdot)$ function has to be given in radians. Thus, n satisfies the equations

$$1 - \Phi\left\{2\sqrt{n}\left(\arcsin \sqrt{\frac{c}{n}} - \arcsin \sqrt{\theta_0}\right)\right\} = \alpha,$$

$$1 - \Phi\left\{2\sqrt{n}\left(\arcsin \sqrt{\frac{c}{n}} - \arcsin \sqrt{\theta_1}\right)\right\} = 1 - \beta. \tag{1.3.14}$$

A solution of n from the equations is

$$n^* = \left\{ \frac{(z_\alpha + z_\beta)}{2(\arcsin \sqrt{\theta_1} - \arcsin \sqrt{\theta_0})} \right\}^2 . \tag{1.3.15}$$

The following result is thus established:

Theorem 1.3.2 *An approximate sample size n required to give a power of at least $1 - \beta$ at the alternative $\theta = \theta_1$ $(>\theta_0)$ to an α-level one-sided test of H_0: $\theta = \theta_0$ is $[n^*] + 1$, where n^* is given by (1.3.15).*

Example 1.3.2.1 A pharmaceutical company claims that its new pain reliever is more than 90% efficient in providing relief. It is decided to conduct a 0.05 level one-sided test of H_0: $\theta = .90$ against the alternative H_A: $\theta > .90$. It is desirable that the test has a power of 0.90, when $\theta = 0.95$. Thus, $\alpha = 0.05$, $1 - \beta = 0.90$, $\theta_0 = 0.90$, $\theta_1 = 0.95$, and

$$n^* = \left(\frac{1.645 + 1.282}{2.6906 - 2.4981} \right)^2 = 231.2.$$

The required sample size is thus $n = 232$.

1.4 Sampling from an Exponential Distribution

Let X be a random variable having a two-parameter (negative) exponential distribution with density function

$$f(x; \gamma, \theta) = \theta^{-1} e^{-(x-\gamma)/\theta} \qquad x > \gamma. \tag{1.4.1}$$

This fact will be denoted by writing $X \sim NE \ (\gamma, \theta)$. Let X_1, $X_2, \ldots, X_n$ be a random sample on X. Let Y_1 be the sample minimum, and $Y_2 = \Sigma_{i=1}^n \ (X_i - Y_1)$. It is then known that $Y_1 \sim NE(\gamma, \theta/n)$, $(2Y_2/\theta) \sim \chi^2 \ (2(n-1))$, and Y_1 and Y_2 are independent.

In the following sections, estimation and tests of hypotheses of the parameters γ and θ will be considered.

1.4.1   Estimation of γ When θ Is Known

The usual estimator of γ is Y_1. By controlling the absolute error $|Y_1 - \gamma|$, the sample size n will be determined meeting the specification

$$P(|Y_1 - \gamma| \leqslant d) \geqslant 1 - \alpha,$$

that is,

$$P(0 \leqslant Y_1 - \gamma \leqslant d) \geqslant 1 - \alpha. \qquad (1.4.2)$$

The required n here is $[n^*] + 1$, where

$$n^* = \theta \frac{\ln\left(\dfrac{1}{\alpha}\right)}{d}. \qquad (1.4.3)$$

Theorem 1.4.1   *The required sample size n for estimating γ when θ is known satisfying the requirement* (1.4.2) *is* $[n^*] + 1$, *where* n^* *is given by* (1.4.3).

1.4.2   Estimation of γ When θ Is Unknown

As in the case of estimating mean when variance is unknown of a normal distribution, here also a two-stage procedure (*cf.* Desu *et al.*, 1976) is desirable and can be carried out in the following steps:

(i) Take an initial sample of size n_1 and estimate θ by

$$\hat{\theta} = \sum_{i=1}^{n_1} \frac{X_i - Y_1(n_1)}{n_1 - 1}, \qquad (1.4.4)$$

where $Y_1(n_1)$ is the minimum of this sample.

(ii) Calculate c from the formula

$$c = (n_1 - 1) \frac{\alpha^{1/(1-n_1)} - 1}{d}. \qquad (1.4.5)$$

(iii) Let $n = \max\{n_1, [c\hat{\theta}] + 1\}$.

(iv) Take $n - n_1$ additional observations, if necessary. Let $Y_1(n)$ be the smallest observation for the combined sample of size n.

The estimator $Y_1(n)$ then meets the specification (1.4.2).

1.4.3 Estimation of θ

An unbiased estimator of θ is

$$\hat\theta = \sum_{i=1}^{n} \frac{X_i - \gamma}{n}, \qquad \text{if } \gamma \text{ is known;}$$

$$= \sum_{i=1}^{n} \frac{X_i - Y_1}{n}, \qquad \text{if } \gamma \text{ is unknown.} \tag{1.4.6}$$

Let $f = 2n$ or $2(n - 1)$ depending on whether γ is known or not. It is known that $f\hat\theta/\theta \sim \chi^2(f)$.

In this case, one can determine f by controlling the relative error. Thus, f will be found to satisfy the requirement

$$P\left(\left|\frac{\hat\theta - \theta}{\theta}\right| \leq r\right) \geq 1 - \alpha. \tag{1.4.7}$$

This implies that

$$P\left((1 - r)f \leq \frac{f\hat\theta}{\theta} \leq (1 + r)f\right) \geq 1 - \alpha, \tag{1.4.8}$$

which can be written as

$$G_f((1 + r)f) - G_f((1 - r)f) \geq 1 - \alpha. \tag{1.4.9}$$

Now,

Theorem 1.4.2 *The sample size n required for estimating θ by $\hat\theta$ meeting the specification (1.4.7) is $f/2$ or $f/2 + 1$ depending on whether γ is known or not, where f is the smallest even positive integer satisfying (1.4.9).*

Note that Equation (1.4.9) is similar to Equation (1.2.11). Arguing in the same manner as in Section 1.2.3, the following theorem can be proven:

Theorem 1.4.3 *An approximation to the sample size n satisfying criterion* (1.4.7) *is* $[f^*/2] + 1$ *or* $[f^*/2] + 2$ *according as* γ *is known or unknown, where*

$$f^* = \frac{2z_{\alpha/2}^2}{r^2}. \qquad (1.4.10)$$

1.4.4 Tests of Hypothesis about θ

Continuing the notation of Section 1.4.3, the critical region of an α-level test of $H_0: \theta = \theta_0$ against the one-sided alternative $H_A: \theta > \theta_0$ is

$$\hat{\theta} > \frac{\theta_0}{f} \chi_{1-\alpha}^2(f). \qquad (1.4.11)$$

To have a power of $1 - \beta$ at $\theta = \theta_1$ ($>\theta_0$), the required sample size n must satisfy the condition

$$P(\hat{\theta} > \frac{\theta_0}{f} \chi_{1-\alpha}^2(f) \,|\, \theta = \theta_1) = 1 - \beta. \qquad (1.4.12)$$

Thus, f must satisfy the relation

$$\theta_0 \chi_{1-\alpha}^2(f) = \theta_1 \chi_\beta^2(f). \qquad (1.4.13)$$

This result can be stated as

Theorem 1.4.4 *The sample size n required to give a power of* $1 - \beta$ *at the alternative* $\theta = \theta_1$ ($>\theta_0$) *to an* α-level one-sided test of $H_0: \theta = \theta_0$ *is* $f/2$ *or* $f/2 + 1$ *depending on whether* γ *is known or unknown, where f is the smallest even positive integer satisfying* (1.4.13).

Approximating χ^2 percentiles with standard normal percentiles as an Section 1.2.6, the following result can be established.

Theorem 1.4.5 *An approximate sample size n required to give a power of* $1 - \beta$ *at the alternative* $\theta = \theta_1$ ($>\theta_0$) *to an* α-level *one-sided test H_0: $\theta = \theta_0$ is $[f^*/2] + 1$ or $[f^*/2] + 2$, where*

$$f^* = \frac{1}{2}\left(\frac{\lambda z_\alpha + z_\beta}{\lambda - 1}\right)^2 + \frac{1}{2}, \qquad (1.4.14)$$

λ^2 *being* θ_0/θ_1.

The case of negative exponential distribution with a single parameter θ corresponds to known $\gamma = 0$, and the appropriate results of this section are applicable. The case of first t failures only will be discussed in the Problems 3 and 4 at the end of this monograph.

1.5 Bayesian Approach to Estimation

The entire discussion up until now used classical methods where probabilities are to be interpreted as relative frequencies. When individuals have prior beliefs about the parameter values, they may want to revise these beliefs in light of the information provided by the samples. This way of making inference, where probabilities are usually interpreted as personal beliefs, leads us to the following Bayesian approach to the sample size problem in the context of estimation of parameters.

Here, a general framework needed to formulate the problem will be given. This discussion is along the lines of Guttman *et al.* (1982).

A random sample of size n on X, with probability function or probability density function $f(x \mid \underline{\theta})$ will be observed to provide information on $\underline{\theta}$. The parameter $\underline{\theta}$ is a continuous variable taking values in the set Θ. Prior beliefs can be stated in the form of a prior distribution defined by $p(\underline{\theta})$. Let $\underline{X}$ be a random sample on X. Given $\underline{X}$, the posterior probability density function of $\underline{\theta}$ is

$$\pi^*(\underline{\theta}|\underline{X}) = \frac{p(\underline{\theta})L(\underline{\theta}|\underline{X})}{\int_\Theta p(\underline{\theta})L(\underline{\theta}|\underline{X})\,d\underline{\theta}}, \qquad (1.5.1)$$

where L is the likelihood function of the sample. The posterior distribution summarizes all the information that one has about the values of $\underline{\theta}$.

Now we concentrate on one component θ_1 of $\underline{\theta}$. The marginal posterior density of θ_1 is $\pi(\theta_1|\underline{X})$, and this will be used for inference purposes. The mean of this posterior distribution is used as a point estimate and is called the Bayes estimate of θ_1. If point estimation is the objective of the experimenter, the sample size determination may be formulated as follows.

Let $\hat{\theta}_1(\underline{X};p)$ be the posterior mean of θ_1, based on the prior p and the random sample $\underline{X}$. One may want to choose the sample size n so that

$$P(|\theta_1 - \hat{\theta}_1(\underline{X};p)| \le d|\underline{X}) \ge 1 - \alpha, \tag{1.5.2}$$

for chosen positive d and α. It may be noted that choosing n in this manner amounts to ensuring that $\hat{\theta}_1(\underline{X};p) \pm d$ is a posterior interval of level $1 - \alpha$ for θ_1. In other words, one wants to ensure that an interval of specified width $2d$ around the Bayes estimate is a posterior interval of specified level $1 - \alpha$.

Alternatively, one may want to choose the sample size n so that

$$P\left(\left|\frac{\theta_1 - \hat{\theta}_1(\underline{X};p)}{\hat{\theta}_1(\underline{X};p)}\right| \le r|\underline{X}\right) \ge 1 - \alpha, \tag{1.5.3}$$

for chosen positive $r(<1)$ and α. When n is chosen in this manner, the interval

$$([1 - r]\hat{\theta}_1(\underline{X};p), \quad [1 + r]\hat{\theta}_1(\underline{X};p))$$

is a posterior interval of level $1 - \alpha$ for θ_1. This interval has the property

$$\frac{\text{Upper end point}}{\text{Middle point}} - 1 = 1 - \frac{\text{Lower end point}}{\text{Middle point}} = r.$$

The results for the same probability model differ from one prior to another. So no detailed discussion will be given. However, some cases, where one can determine the sample size n using one of the previously mentioned requirements, are included in the problems and supplements section (Problems 14–16). For a good review of the concepts and results about the posterior distributions, one may refer to Chapter 9A of Guttman et al. (1982). For other decision theory sample size problems, the interested reader is referred to DeGroot (1970, Chapters 8 and 9).

Chapter 2　　　　　Two-Sample Problems

2.1　Introduction

Let X and Y be two independent random variables with distribution functions $F(x)$ and $G(y)$, respectively. One may compare these distributions or the appropriate parameters of these based on two independent random samples on X and Y. For example, one may draw inferences on the difference of their means, $E(X) - E(Y)$, or on the ratio of their variances, $\mathrm{Var}\,(X)/\mathrm{Var}(Y)$.

The problem of determining the appropriate sizes of the samples to draw various inferences meeting certain specifications will be discussed in this chapter.

2.2　Sampling from Two Normal Distributions

Let $X \sim N(\mu_1, \sigma_1^2)$ and $Y \sim N(\mu_2, \sigma_2^2)$. A random sample $X_1, X_2, \ldots, X_{n_1}$ of size n_1 is taken on X and an independent random sample $Y_1, Y_2, \ldots, Y_{n_2}$ of size n_2 is taken on Y. Let

$$\bar{X} = \sum_{i=1}^{n_1} \frac{X_i}{n_1}, \qquad s_1^2 = \sum_{i=1}^{n_1} \frac{(X_i - \bar{X})^2}{n_1 - 1},$$

$$\bar{Y} = \sum_{j=1}^{n_2} \frac{Y_j}{n_2}, \qquad s_2^2 = \sum_{j=1}^{n_2} \frac{(Y_j - \bar{Y})^2}{n_2 - 1}. \tag{2.2.1}$$

The estimation and tests of hypotheses about the parameters $\mu_1 - \mu_2$ and σ_1^2/σ_2^2 will be considered in the following sections.

2.2.1   Estimation of $\mu_1 - \mu_2$ When σ_1^2, σ_2^2 Are Known

The usual unbiased estimator of $\mu_1 - \mu_2$ is $\bar{X} - \bar{Y}$ and is distributed as $N(\mu_1 - \mu_2, \sigma_1^2/n_1 + \sigma_2^2/n_2)$. By controlling the absolute error of the estimator with high probability, one wants to determine n_1 and n_2 such that

$$P(|(\bar{X} - \bar{Y}) - (\mu_1 - \mu_2)| \leq d) \geq 1 - \alpha, \tag{2.2.2}$$

where d and α are prechosen positive constants. To get unique solution to n_1 and n_2 satisfying the condition (2.2.2), there is a need to have a second condition. Taking the cost and other factors into consideration, one usually imposes a condition $n_1 = kn_2$ for a suitable k. In most problems, however, the special case of $k = 1$ (equal sample sizes) will be considered. The following result about n_1 and n_2 easily follows.

Theorem 2.2.1   *The appropriate sizes n_1 and n_2 of the two samples meeting the specification (2.2.2) are $n_1 = [kn^*] + 1$, $n_2 = [n^*] + 1$, where*

$$n^* = \frac{\{(\sigma_1^2/k) + \sigma_2^2\}z_{\alpha/2}^2}{d^2}. \tag{2.2.3}$$

When $\sigma_1^2 = \sigma_2^2$ ($= \sigma^2$, say) it is customary to take $n_1 = n_2$ irrespective of the cost differential and in that case

$$n_1 = n_2 = \left[\frac{2\sigma^2 z_{\alpha/2}^2}{d^2}\right] + 1. \tag{2.2.4}$$

Example 2.2.1.1   A physician is interested in estimating the difference in mean pulse rates for men and women to within one beat with a probability of at least 0.95. Suppose that the common standard deviation of the pulse rates among men and women is two beats. Now from Equation (2.2.4),

$$n_1 = n_2 = \left[\frac{2(2^2)(1.96^2)}{1^2} \right] + 1 = 31.$$

Thus, he should randomly select 31 men and 31 women for this study.

2.2.2   Estimation of $\mu_1 - \mu_2$ When $\sigma_1^2 = \sigma_2^2$ and the Common Variance Is Unknown

In this case, it is necessary to use a two-stage procedure to meet the specification given in (2.2.2), as described:

(i) Take initial samples of size $n_1(\geq 2)$ from each of the two populations.

(ii) Let s_p^2 be the pooled variance of these two samples.

(iii) Calculate $c = \dfrac{1}{2} \left\{ \dfrac{d}{t_{\alpha/2}(2n_1 - 2)} \right\}^2$.

(iv) Let $n = \max\{n_1, \left[\dfrac{s_p^2}{c} \right] + 1\}$.

(v) Take $n - n_1$ additional observations, if necessary, from each of the two populations. Let $\bar{X}$ and $\bar{Y}$ be the two sample means based on all n observations.

Then $\bar{X} - \bar{Y}$ as an estimator of $\mu_1 - \mu_2$ meets the requirement (2.2.2). See Problem 8 for mathematical details of this procedure.

In Example 2.2.1.1, suppose that the population standard deviation for the pulse rate distribution for men and women is unknown. Independent random samples of size five can be taken from each of the two populations and let the pulse rates be as given below:

$$
\begin{array}{ll}
\text{men} & 81,\ 79,\ 78,\ 82,\ 80; \\
\text{women} & 77,\ 76,\ 78,\ 79,\ 77.
\end{array}
$$

Clearly, $s_p^2 = 1.9$. Now,

$$c = \frac{1}{2} \left\{ \frac{1}{2.306} \right\}^2 = 0.094$$

$$n = \max \left\{ 5, \left[\frac{1.9}{0.094} \right] + 1 \right\} = 21.$$

An additional sample of 16 subjects has to be taken from each sex group.

2.2.3 Estimation of $\mu_1 - \mu_2$ When $\sigma_1^2 \neq \sigma_2^2$ and the Variances Are Unknown

Koopmans and Qualls (1971) suggested a two-stage procedure for this case to meet the requirement (2.2.2), which is described next:

  (i) Take first stage independent samples of sizes n_{11} (≥ 2) and n_{21} (≥ 2) from the two populations. Let s_{11}^2 and s_{21}^2 be the two variances of these samples.
 (ii) Calculate upper $100\sqrt{1 - (\alpha/2)}\%$ confidence limits on σ_1^2 and σ_2^2. Let σ_{1u}^2 and σ_{2u}^2 be these limits.
(iii) Let $C = c_0 + c_1 n_{12} + c_2 n_{22}$ be the linear cost function, where n_{12} and n_{22} are the second stage sample sizes from the two populations. The solutions of n_{12} and n_{22} that minimize C subject to

$$z_{\alpha/4} \sqrt{\frac{\sigma_{1u}^2}{n_{12}} + \frac{\sigma_{2u}^2}{n_{22}}} \leq d, \qquad (2.2.5)$$

are

$$n_{12} = \left[\frac{z_{\alpha/4}^2}{d^2} \frac{\sqrt{c_1}\,\sigma_{1u} + \sqrt{c_2}\,\sigma_{2u}}{\sqrt{c_1}} \sigma_{1u} \right] + 1,$$

$$n_{22} = \left[\frac{z_{\alpha/4}^2}{d^2} \frac{\sqrt{c_1}\,\sigma_{1u} + \sqrt{c_2}\,\sigma_{2u}}{\sqrt{c_2}} \sigma_{2u} \right] + 1. \qquad (2.2.6)$$

 (iv) Let $\bar{X}_2$ and $\bar{Y}_2$ be the two sample means for the second stage samples.

Then, $\bar{X}_2 - \bar{Y}_2$ as an estimator of $\mu_1 - \mu_2$ satisfies the requirement of (2.2.2).

It may be noted that the required estimator is based on the means of the second-stage samples only in this procedure.

2.2.4 Estimation of Variance Ratio, σ_1^2/σ_2^2

Let

$$\hat{\sigma}_1^2 = \sum_{i=1}^{n_1} \frac{(X_i - \mu_1)^2}{n_1}, \qquad \text{if } \mu_1 \text{ is known,}$$

$$= s_1^2, \qquad \text{if } \mu_1 \text{ is unknown;}$$

$$\hat{\sigma}_2^2 = \sum_{j=1}^{n_2} \frac{(Y_j - \mu_2)^2}{n_2}, \qquad \text{if } \mu_2 \text{ is known;}$$

$$= s_2^2, \qquad \text{if } \mu_2 \text{ is unknown.}$$

(2.2.7)

Let $\nu_i = n_i$ or $n_i - 1$ depending on whether μ_i is known or not, $i = 1,2$. A commonly used estimator of σ_1^2/σ_2^2 is $\hat{\sigma}_1^2/\hat{\sigma}_2^2$ and $(\hat{\sigma}_1^2/\hat{\sigma}_2^2)/(\sigma_1^2/\sigma_2^2)$ is distributed as a F-distribution with ν_1 and ν_2 degrees of freedom. This fact is denoted by writing $(\hat{\sigma}_1^2/\hat{\sigma}_2^2)/(\sigma_1^2/\sigma_2^2) \sim F(\nu_1, \nu_2)$. In this case, one controls the relative error by specifying the condition

$$P\left(\frac{\left| \left(\dfrac{\hat{\sigma}_1^2}{\hat{\sigma}_2^2}\right) - \left(\dfrac{\sigma_1^2}{\sigma_2^2}\right) \right|}{\left(\dfrac{\sigma_1^2}{\sigma_2^2}\right)} \le r \right) \ge 1 - \alpha. \tag{2.2.8}$$

This simplifies to

$$P(1 - r \le F(\nu_1, \nu_2) \le 1 + r) \ge 1 - \alpha. \tag{2.2.9}$$

Let $H_{\nu_1,\nu_2}(\cdot)$ be the distribution function of $F(\nu_1, \nu_2)$ variable. Then (2.2.9) can be rewritten as

$$H_{\nu_1,\nu_2}(1 + r) - H_{\nu_1,\nu_2}(1 - r) \ge 1 - \alpha. \tag{2.2.10}$$

Thus,

Theorem 2.2.2   *The sample sizes needed to meet the specification*
(2.2.8) are solutions satisfying the equation (2.2.10).

One may get an approximate solution of (2.2.10) by noting that
the asymptotic distribution of

$$Z = \frac{\frac{1}{2}\ln F(\nu_1, \nu_2) - \frac{1}{2}\left(\frac{1}{\nu_2} - \frac{1}{\nu_1}\right)}{\sqrt{\frac{1}{2}\left(\frac{1}{\nu_1} + \frac{1}{\nu_2}\right)}}, \qquad (2.2.11)$$

is $N(0, 1)$. Assuming $1/\nu_2$ is nearly equal to $1/\nu_1$, (2.2.10) can be
asymptotically approximated to the inequality

$$\Phi\left\{\frac{\frac{1}{2}\ln(1+r)}{\sqrt{\frac{1}{2}\left(\frac{1}{\nu_1} + \frac{1}{\nu_2}\right)}}\right\} - \Phi\left\{\frac{\frac{1}{2}\ln(1-r)}{\sqrt{\frac{1}{2}\left(\frac{1}{\nu_1} + \frac{1}{\nu_2}\right)}}\right\} \geq 1 - \alpha \quad (2.2.12)$$

The argument of the second term on the left-hand side of the
equation tends to negative infinity, and the inequality reduces to

$$\Phi\left\{\frac{\frac{1}{2}\ln(1+r)}{\sqrt{\frac{1}{2}\left(\frac{1}{\nu_1} + \frac{1}{\nu_2}\right)}}\right\} \geq 1 - \alpha. \qquad (2.2.13)$$

A solution to the inequality is

$$\frac{\frac{1}{2}\ln(1+r)}{\sqrt{\frac{1}{2}\left(\frac{1}{\nu_1} + \frac{1}{\nu_2}\right)}} = z_\alpha, \qquad (2.2.14)$$

and this is also an approximate solution to the equation (2.2.10).

Now

$$\frac{1}{\nu_1} + \frac{1}{\nu_2} = \frac{1}{2}\left\{\frac{\ln(1+r)}{z_\alpha}\right\}^2 = \tau, \text{say.} \qquad (2.2.15)$$

To get an unique solution for ν_1 and ν_2, consider minimizing the linear cost function, $C = c_0 + c_1 n_1 + c_2 n_2$, which will be conveniently written as

$$c^* = c_1\nu_1 + c_2\nu_2, \qquad (2.2.16)$$

for an appropriate c^*. Using the Cauchy-Schwartz inequality

$$(c_1\nu_1 + c_2\nu_2) \times \left(\frac{1}{\nu_1} + \frac{1}{\nu_2}\right) \geq (\sqrt{c_1} + \sqrt{c_2})^2, \qquad (2.2.17)$$

one gets minimum c^*, when

$$\nu_1^* = \frac{1}{\tau}\left(1 + \sqrt{\frac{c_2}{c_1}}\right), \qquad \nu_2^* = \frac{1}{\tau}\left(1 + \sqrt{\frac{c_1}{c_2}}\right), \qquad (2.2.18)$$

are solutions of ν_1 and ν_2. Thus,

Theorem 2.2.3 *The approximate sample sizes n_i needed to meet the specification (2.2.8) minimizing the cost function (2.2.16) are $[\nu_i^*] + 1$ or $[\nu_i^*] + 2$, according as μ_i is known or not, where ν_i^* are given in (2.2.18) for $i = 1, 2$.*
  When $c_1 = c_2$, (2.2.18) simplifies to

$$\nu_1^* = \nu_2^* = \frac{2}{\tau} = 4\left\{\frac{z_\alpha}{\ln(1+r)}\right\}^2. \qquad (2.2.19)$$

Example 2.2.3.1 An industrial engineer is interested in estimating the ratio of the variances of the thicknesses of microchips produced by two companies with a maximum relative error rate of 20% with a probability of 95%. Here, $r = 0.2$, $\alpha = 0.05$, and one can assume $c_1 = c_2$ and $\nu_1 = \nu_2$. From Equation (2.2.19),

$$\nu_1^* = \nu_2^* = 4\left\{\frac{1.645}{\ln(1 + 0.2)}\right\}^2 = 325.62$$

The sample sizes needed from each population are 326 or 327 depending on whether the means are known or not.

2.2.5 Tests of Hypothesis about μ_1 and μ_2 When σ_1^2 and σ_2^2 Are Known

Consider an α-level test of H_0: $\mu_1 = \mu_2$ against the one-sided alternative H_A: $\mu_1 > \mu_2$. The critical region for this test is

$$\bar{X} - \bar{Y} > z_\alpha \sqrt{\frac{\sigma_1^2}{n_1} + \frac{\sigma_2^2}{n_2}}. \qquad (2.2.20)$$

If a power of $1 - \beta$ is required at $\mu_1 - \mu_2 = \delta(>0)$, n_1 and n_2 must satisfy

$$P\left(\bar{X} - \bar{Y} > z_\alpha \sqrt{\frac{\sigma_1^2}{n_1} + \frac{\sigma_2^2}{n_2}} \,\middle|\, \mu_1 - \mu_2 = \delta\right) = 1 - \beta. \quad (2.2.21)$$

Thus,

$$\frac{\sigma_1^2}{n_1} + \frac{\sigma_2^2}{n_2} = \left\{\frac{\delta}{z_\alpha + z_\beta}\right\}^2. \qquad (2.2.22)$$

To get an unique solution to (2.2.22), one takes $n_1 = kn_2$ and gets

$$n_2^* = \frac{\dfrac{\sigma_1^2}{k} + \sigma_2^2}{\left(\dfrac{\delta}{z_\alpha + z_\beta}\right)^2}, \qquad n_1^* = kn_2^*. \qquad (2.2.23)$$

Thus,

Theorem 2.2.4   *The sample sizes n_1 and n_2 needed to give a power of $1 - \beta$ at the alternative $\mu_1 - \mu_2 = \delta(>0)$ for an α-level one-sided test of H_0: $\mu_1 = \mu_2$ are $[kn_2^*] + 1$ and $[n_2^*] + 1$, respectively, where n_2^* is given as in (2.2.23).*

A similar result can be obtained for a two-sided test using $\alpha/2$ for α in (2.2.23).

2.2.6 Tests of Hypothesis about μ_1 and μ_2 When $\sigma_1^2 = \sigma_2^2$ and the Common Variance Is Unknown

Consider an α-level test of $H_0: \mu_1 = \mu_2$ against the one-sided alternative $H_A: \mu_1 > \mu_2$, based on n observations from each population. The critical region of the usual t-test is

$$\bar{X} - \bar{Y} > t_\alpha(\nu)s_p \sqrt{\frac{2}{n}}, \tag{2.2.24}$$

where $\nu = 2(n-1)$ and s_p^2 is the pooled variance. By requiring a power of $1 - \beta$ when $\mu_1 - \mu_2 = \delta$, one gets the condition

$$P\left(\bar{X} - \bar{Y} > t_\alpha(\nu)s_p \sqrt{\frac{2}{n}} \middle| \delta\right) = 1 - \beta, \tag{2.2.25}$$

to solve for n. Then, (2.2.25) reduces to

$$P\left(T(\nu) > t_\alpha(\nu) - \frac{\delta}{s_p\sqrt{\frac{2}{n}}}\right) = 1 - \beta. \tag{2.2.26}$$

Approximating s_p by σ, Equation (2.2.26) gives

$$n = \frac{\nu}{2} + 1 \doteq 2\left(\frac{\sigma}{\delta}\right)^2 \{t_\alpha(\nu) + t_\beta(\nu)\}^2. \tag{2.2.27}$$

This equation should be solved iteratively for ν. Thus,

Theorem 2.2.5   *The sample sizes $n_1 = n_2 = n$ needed to give a power of $1 - \beta$ when $\mu_1 - \mu_2 = \delta$ (>0) for a one-sided α-level test of $H_0: \mu_1 = \mu_2$ is $(\nu/2 + 1)$, where ν is the smallest positive even integer satisfying* (2.2.27).

The result of Theorem 2.2.5 was suggested by Cochran and Cox (1957). One needs to use an independent estimate of σ or an upper bound, σ_u, of σ to solve (2.2.27).

Example 2.2.5.1   A physician is interested in ascertaining whether the average pulse rate for males is more than that of females

using a 0.05 level one-sided test. He wants to detect the difference with a probability of 0.9, when the difference in the average pulse rates is three. It is known that the common standard
deviation of pulse rates for the groups does not exceed two. Here,
$\sigma_u = 2$, $\delta = 3$, $\alpha = 0.05$, and $\beta = 0.1$. Now,

$$\frac{\nu}{2} + 1 = 2\left(\frac{2}{3}\right)^2 \{t_{0.05}(\nu) + t_{0.10}(\nu)\}^2.$$

Initially, replacing the Student's t percentiles with standard normal percentiles, one gets

$$\nu = 2\left\{\frac{8}{9}(1.645 + 1.282)^2 - 1\right\}$$

$$= 13.2.$$

Now take $\nu = 14$ and find Student's t percentiles to get

$$\nu = 2\left\{\frac{8}{9}(1.761 + 1.345)^2 - 1\right\}$$

$$= 15.1.$$

For the next approximation take $\nu = 16$ and find Student's t
percentiles to get

$$\nu = 2\left\{\frac{8}{9}(1.746 + 1.337)^2 - 1\right\} = 14.6.$$

Thus, ν will be taken as 16, and the required common sample size
is $n_1 = n_2 = 9$.

An alternative approach to this problem is a two-stage procedure described next:

(i) Take initial samples of size n_1 from each of the two
populations.

(ii) Let s_{p1}^2 be the pooled variance of these two samples

(iii) Calculate $c = (\delta^2/2)\{t_\alpha(\nu) + t_\beta(\nu)\}^{-2}$, where $\nu = 2(n_1 - 1)$.

(iv) Let $n = \max\{n_1, [s_{p1}^2/c] + 1\}$.

(v) Take $n - n_1$ additional observations, if necessary, from each of the two populations. Let $\bar{X}$ and $\bar{Y}$ be the two sample means based on all n observations.

Then the critical region of an α-level one-sided test of H_0: $\mu_1 = \mu_2$ that gives a power of $1 - \beta$ at $\mu_1 - \mu_2 = \delta(>0)$ is

$$\bar{X} - \bar{Y} > \{t_\alpha(2n_1 - 2)\}s_{p_1}\sqrt{\frac{2}{n}}. \qquad (2.2.28)$$

For a two-sided test, a similar solution using $\alpha/2$ for α is available. For mathematical details of this procedure, refer to Problem 14 at the end of this monograph.

2.2.7 Tests of Hypothesis about μ_1 and μ_2 When $\sigma_1^2 \neq \sigma_2^2$ and the Variances Are Unknown

Stein's (1945) two-stage procedure was extended to a two-sample case by Chapman (1950). This procedure uses a test statistic, which is the difference of two independent Student's t variables with the same degrees of freedom. Let $T_1(\nu)$ and $T_2(\nu)$ be two independent t variables with ν degrees of freedom each. Let

$$M(\nu) = T_1(\nu) - T_2(\nu). \qquad (2.2.29)$$

A short table of the distribution of this variable is given in Chapman (1950). Let $m_p(\nu)$ be the upper $100p$ percentile of $M(\nu)$.

Consider the problem of testing H_0: $\mu_1 = \mu_2$ against H_A: $\mu_1 > \mu_2$ at α level of significance. Further, let it be desired to get a power of $1 - \beta$ when $\mu_1 - \mu_2 = \delta(>0)$. The following procedure gives the required test:

(i) Take two independent samples each of size n_0 from the two populations. Let s_{11}^2 and s_{21}^2 be their sample variances.

(ii) Calculate $c = \delta^2\{m_\alpha(n_0 - 1) + m_\beta(n_0 - 1)\}^{-2}$.

(iii) Let $n_i = \max(n_0 + 1, [s_{i1}^2/c] + 1)$, $\quad i = 1, 2$.

(iv) Take $n_i - n_0$ additional observations from the ith population, $i = 1, 2$.

(v) Let

$$b_i = \frac{1}{n_i} \left\{ 1 + \sqrt{\frac{n_0(n_i c - s_{i1}^2)}{(n_i - n_0)s_{i1}^2}} \right\},$$

$$a_i = \frac{1 - (n_i - n_0)b_i}{n_0}, \qquad i = 1, 2.$$

(2.2.30)

(vi) Let X_i, $i = 1, 2, \ldots, n_1$ be the sample on X and let Y_j, $j = 1, 2, \ldots, n_2$ be the sample on Y. Let

$$T_1 = \frac{\sum_{i=1}^{n_0} a_1 X_i + \sum_{i=n_0+1}^{n_1} b_1 X_i}{\sqrt{c}},$$

$$T_2 = \frac{\sum_{j=1}^{n_0} a_2 Y_j + \sum_{j=n_0+1}^{n_2} b_2 Y_j}{\sqrt{c}}.$$

(2.2.31)

The critical region for the required test is

$$T_1 - T_2 > m_\alpha(n_0 - 1). \tag{2.2.32}$$

The two-sided test can be similarly obtained.

2.2.8 Tests of Hypothesis about σ_1^2 and σ_2^2

Continuing the notation of Section 2.2.4, consider an α-level test of H_0: $\sigma_1^2 = \sigma_2^2$ against the one-sided alternative H_A: $\sigma_1^2 > \sigma_2^2$. The critical region of the usual test is

$$\frac{\hat{\sigma}_1^2}{\hat{\sigma}_2^2} > F_{1-\alpha}(\nu_1, \nu_2), \tag{2.2.33}$$

where $F_{1-\alpha}(\nu_1, \nu_2)$ are $(1 - \alpha) 100$ percentile points of $F(\nu_1, \nu_2)$ distribution. By requiring a power of $1 - \beta$, when $\sigma_1^2/\sigma_2^2 = \lambda^2$ (>1), one gets the specification

$$P\left(\frac{\hat{\sigma}_1^2}{\hat{\sigma}_2^2} > F_{1-\alpha}(\nu_1, \nu_2) \,\middle|\, \frac{\sigma_1^2}{\sigma_2^2} = \lambda^2 \right) = 1 - \beta, \tag{2.2.34}$$

and the sample sizes are determined satisfying it. This condition leads to

$$P(F(\nu_1, \nu_2) > \frac{1}{\lambda^2} F_{1-\alpha}(\nu_1, \nu_2)) = 1 - \beta. \tag{2.2.35}$$

Thus,

$$F_{1-\alpha}(\nu_1, \nu_2) = \lambda^2 F_\beta(\nu_1, \nu_2). \qquad (2.2.36)$$

This equation should be solved for ν_1 and ν_2, and the following result is obtained.

Theorem 2.2.6 *The sample sizes needed to give a power of $1 - \beta$ when $\sigma_1^2/\sigma_2^2 = \lambda^2(>1)$ for an α-level one-sided test of H_0: $\sigma_1^2 = \sigma_2^2$ are solutions satisfying the Equation (2.2.36).*

The solutions of (2.2.36) need not be unique. One gets an unique solution by imposing a condition such as $\nu_1 = k\nu_2$, or more specifically, $\nu_1 = \nu_2$.

By approximating the percentiles of F distribution with the percentiles of a standard normal distribution (*cf.* Section 2.2.4) and putting $\nu_1 = k\nu_2$, the solutions of ν_1 and ν_2 are given by

$$\nu_1^* = k\nu_2^*, \ \nu_2^* = \left(\frac{k+1}{2k}\right)\left(\frac{z_\alpha + z_\beta}{\ln \lambda}\right)^2. \qquad (2.2.37)$$

Thus, the following is established.

Theorem 2.2.7 *The approximate sample sizes n_1 and n_2 required to give a power of $1 - \beta$ when $\sigma_1^2/\sigma_2^2 = \lambda^2(>1)$ for an α-level one-sided test of H_0: $\sigma_1^2 = \sigma_2^2$ such that $\nu_1 = k\nu_2$ are $[\nu_i^*] + 1$ or $[\nu_i^*] + 2$ according as μ_is are known or not, where ν_i^* are given in (2.2.37) for $i = 1, 2$.*

Example 2.2.7.1 Let σ_1^2 and σ_2^2 be the population variances of thicknesses of microchips manufactured by two companies, and an industrial engineer is interested in testing the null hypothesis H_0: $\sigma_1^2 = \sigma_2^2$ against the one-sided alternative hypothesis $\sigma_1^2 > \sigma_2^2$ using a 0.05 level test. Furthermore, he wants to detect the difference with a probability 0.90 when $\sigma_1^2/\sigma_2^2 = 2.25$. Further, he assumes $\nu_1 = \nu_2$. Then from Equation (2.2.37)

$$\nu_1^* = \nu_2^* = \left(\frac{1.645 + 1.282}{\ln 2.25}\right)^2 = 13.03.$$

The required common sample size is 14 or 15 depending on whether μ_is are known or unknown.

2.3   Sampling from Two Bernoulli Distributions

Let X have a Bernoulli distribution with θ_1 as probability of success and let Y also have a Bernoulli distribution with θ_2 as success probability. Based on a random sample $X_1, X_2, \ldots, X_{n_1}$ on X and an independent random sample $Y_1, Y_2, \ldots, Y_{n_2}$ on Y, one may like to test the null hypothesis $H_0: \theta_1 = \theta_2$ against the one-sided alternative $H_A: \theta_1 > \theta_2$ using an α level of significance. It will be required that this test has a power of $1 - \beta$ at $\theta_1 - \theta_2 = \delta(>0)$.

Let $\hat{\theta}_1 = \Sigma_{i=1}^{n_1} X_i / n_1$, and $\hat{\theta}_2 = \Sigma_{j=1}^{n_2} Y_j / n_2$. An approximate α-level test has the critical region

$$\{\text{arcsin } \sqrt{\hat{\theta}_1} - \text{arcsin } \sqrt{\hat{\theta}_2}\} > \frac{z_\alpha}{2} \sqrt{\frac{1}{n_1} + \frac{1}{n_2}}. \qquad (2.3.1)$$

The power requirement is

$$P\left(\text{arcsin } \sqrt{\hat{\theta}_1} - \text{arcsin } \sqrt{\hat{\theta}_2} > \frac{z_\alpha}{2} \sqrt{\frac{1}{n_1} + \frac{1}{n_2}} \,\middle|\, \theta_1 - \theta_2 = \delta\right) = 1 - \beta,$$

$$(2.3.2)$$

and this requires sample sizes n_1 and n_2 to satisfy

$$\frac{1}{n_1} + \frac{1}{n_2} = \left\{\frac{2\delta^*}{z_\alpha + z_\beta}\right\}^2, \qquad (2.3.3)$$

where $\delta^* = \text{arcsin } \sqrt{\theta_1} - \text{arcsin } \sqrt{\theta_2}$. When θ_1 and θ_2 are individually given, Equation (2.3.3) can be solved for n_1 and n_2 uniquely using a side condition such as $n_1 = kn_2$. Since θ_1 and θ_2 are not individually specified, however, one takes the least favorable configuration and uses

$$\delta_{\min}^* = \text{arcsin } \sqrt{0.50 + \frac{\delta}{2}} - \text{arcsin } \sqrt{0.50 - \frac{\delta}{2}} \qquad (2.3.4)$$

for δ^* in (2.3.3). To get an unique solution for (2.3.3), set $n_1 = kn_2$. Then one gets the solutions

$$n_1^* = kn_2^*, \qquad n_2^* = \frac{(k+1)}{k}\left\{\frac{z_\alpha + z_\beta}{2\delta_{\min}^*}\right\}^2. \qquad (2.3.5)$$

Thus, the following theorem is proved:

Theorem 2.3.1 *The approximate sample size n_1 and n_2 required for an α-level one-sided test of H_0: $\theta_1 = \theta_2$ to give a power of $1 - \beta$ at $\theta_1 - \theta_2 = \delta(>0)$ are $[n_1^*] + 1$, $[n_2^*] + 1$, where n_1^* and n_2^* are given in (2.3.5).*

Example 2.3.1.1 A physician is interested in testing the equality of the response rates of two drugs in curing a disease using a 0.05 level one-sided test. He likes to detect a difference of $\delta = 0.10$ with a probability of 0.90, and he is willing to take $n_1 = n_2$. Now from Equation (2.3.4),

$$\delta_{\min}^* = \arcsin\sqrt{0.50 + 0.05} - \arcsin\sqrt{0.50 - 0.05}$$

$$= 0.8355 - 0.7353 = 0.1002.$$

From Equation (2.3.5)

$$n_1^* = n_2^* = 2\left\{\frac{1.645 + 1.282}{2(0.1002)}\right\}^2 = 426.66.$$

Thus, he should use 427 patients for each drug.

In this case, one could also use Fisher's exact test, and the sample sizes needed in that case are given in Gail and Gart (1973) and Haseman (1978).

2.4 Sampling from Two Exponential Distributions

Let $X \sim NE(0, \theta_1)$ and $Y \sim NE(0, \theta_2)$. A random sample $X_1, X_2, \ldots, X_{n_1}$ of size n_1 is taken on X, and an independent random sample $Y_1, Y_2, \ldots, Y_{n_2}$ of size n_2 is taken on Y. Let $\bar{X}$ and $\bar{Y}$ be the respective means. In the following sections, estimation and tests of hypothesis about θ_1/θ_2 will be considered.

2.4.1 Estimation of θ_1/θ_2

The usual estimator of θ_1/θ_2 is $\bar{X}/\bar{Y}$. In this problem, in controlling the relative error in estimation, one considers the specification

$$P\left(\left|\frac{(\bar{X}/\bar{Y}) - (\theta_1/\theta_2)}{(\theta_1/\theta_2)}\right| \leq r\right) \geq 1 - \alpha. \qquad (2.4.1)$$

Noting that $2\sum_{i=1}^{n_1} X_i/\theta_1 \sim \chi^2(2n_1)$, $2\sum_{j=1}^{n_2} Y_j/\theta_2 \sim \chi^2(2n_2)$, Equation (2.4.1) can be written as

$$P(1-r \leq F(2n_1, 2n_2) \leq 1+r) \geq 1 - \alpha, \qquad (2.4.2)$$

or equivalently,

$$H_{2n_1, 2n_2}(1+r) - H_{2n_1, 2n_2}(1-r) \geq 1 - \alpha. \qquad (2.4.3)$$

The sample sizes have to be calculated iteratively. Thus,

Theorem 2.4.1   *The sample sizes needed to meet the specification (2.4.1) are solutions satisfying the inequality (2.4.3).*

Approximating F percentiles with standard normal percentiles and minimizing the linear cost function $C = c_0 + c_1 n_1 + c_2 n_2$, Equation (2.4.3) can be solved (*cf.* Section 2.2.4), and the following theorem gives the solution.

Theorem 2.4.2   *Approximate sample sizes n_i needed to meet the specification (2.4.1) minimizing the linear cost function $C = c_0 + c_1 n_1 + c_2 n_2$ are $[n_i^*] + 1$, where*

$$n_1^* = \left\{\frac{z_\alpha}{\ln(1+r)}\right\}^2 \left(1 + \sqrt{\frac{c_2}{c_1}}\right).$$

$$n_2^* = \left\{\frac{z_\alpha}{\ln(1+r)}\right\}^2 \left(1 + \sqrt{\frac{c_1}{c_2}}\right). \qquad (2.4.4)$$

2.4.2   *Tests of Hypothesis about θ_1/θ_2*

Consider an α-level test of H_0: $\theta_1/\theta_2 = 1$ against the one-sided alternative H_A: $\theta_1/\theta_2 > 1$. Further, let it be required to have a power of $1 - \beta$ when $\theta_1/\theta_2 = \lambda^2$ (>1). The critical region of the usual test is

$$\frac{\bar{X}}{\bar{Y}} > F_{1-\alpha}(2n_1, 2n_2), \qquad (2.4.5)$$

and the power requirement imposes the condition

$$P\left(\frac{\bar{X}}{\bar{Y}} > F_{1-\alpha}(2n_1, 2n_2) \,\middle|\, \frac{\theta_1}{\theta_2} = \lambda^2\right) = 1 - \beta. \qquad (2.4.6)$$

Thus,

$$F_{1-\alpha}(2n_1, 2n_2) = \lambda^2 F_\beta(2n_1, 2n_2), \qquad (2.4.7)$$

which needs to be solved iteratively. Hence the following:

Theorem 2.4.3 *The sample sizes needed to give a power of $1 - \beta$, where $\theta_1/\theta_2 = \lambda^2(>1)$ for an α-level one-sided test of H_0: $\theta_1/\theta_2 = 1$ are solutions satisfying (2.4.7).*

To get an unique solution of (2.4.7), one imposes a condition such as $n_1 = kn_2$.

By approximating the F percentiles with standard normal percentiles, and imposing the condition $n_1 = kn_2$, the following theorem can be proved.

Theorem 2.4.4 *The approximate sample sizes n_1 and n_2 required to give a power of $1 - \beta$ when $\theta_1/\theta_2 = \lambda^2(>1)$ for an α-level one-sided test of H_0: $\theta_1/\theta_2 = 1$ such that $n_1 = kn_2$ are $[n_i^*] + 1$, $i = 1, 2$, where*

$$n_1^* = kn_2^*, \qquad n_2^* = \frac{(k+1)}{4k}\left\{\frac{z_\alpha + z_\beta}{\ln \lambda}\right\}^2. \qquad (2.4.8)$$

Chapter 3   *k*-Sample Problems

3.1 Introduction

Let X_1, $X_2, \ldots$, X_k be k independent random variables. Let $F_i(x)$ be the distribution function of X_i, $i = 1, 2, \ldots$, k. Further, let $\mu_i = E(X_i)$ and $\sigma_i^2 = \mathrm{var}(X_i)$ for $i = 1, 2, \ldots, k$.

Based on random samples from these populations, one may like to estimate all differences of the form $\mu_i - \mu_j$ for $i \neq j$, $i, j = 1, 2, \ldots, k$. Sometimes one may be interested in comparing μ_1 with each of the other μ_is for $i = 2, 3, \ldots, k$. This is the case when X_1 is the response variable for a control treatment, and $X_2, X_3, \ldots, X_k$ are response variables for $(k - 1)$ active treatments under study. The interest then lies in estimating the $k - 1$ differences, $\mu_i - \mu_1$, for $i = 2, 3, \ldots, k$.

In addition, one may like to test the hypotheses $H_0: F_1 = F_2 = \cdots = F_k$ by controlling the power at certain alternatives of interest. This testing problem will be considered by assuming that F_is belong to some given parametric families.

3.2   Estimation of Difference of Means

The estimation of differences of means without distributional assumption and under normality assumption will be discussed in the following sections.

3.2.1   Estimation without Distributional Assumptions

Let X_{i1}, $X_{i2}, \ldots, X_{in_i}$ be a random sample of size n_i on X_i for $i = 1, 2, \ldots, k$; and, let

$$\bar{X}_{i.} = \sum_{j=1}^{n_i} \frac{X_{ij}}{n_i}. \tag{3.2.1}$$

Then $\bar{X}_{i.} - \bar{X}_{j.}$ is the commonly used unbiased estimator of $\mu_i - \mu_j$. Since there are $k(k-1)/2$ such differences, one may like to estimate all the differences with minimum average variance for the estimators. If A is the average variance of all $k(k-1)/2$ estimators, then

$$A = \frac{2}{k(k-1)} \sum_{\substack{i,j=1 \\ i<j}}^{k} \mathrm{var}(\bar{X}_{i.} - \bar{X}_{j.})$$

$$= \frac{2}{k} \sum_{i=1}^{k} \frac{\sigma_i^2}{n_i}, \tag{3.2.2}$$

and this will be minimized using the constraint on the linear cost function

$$c_0 + \sum_{i=1}^{n} c_i n_i = C. \tag{3.2.3}$$

Using Cauchy-Schwartz inequality, the following result is established:

Theorem 3.2.1   *The sample sizes n_i, which minimize the average variance A of the estimators of the $k(k-1)/2$ differences of the parameters, $\mu_i - \mu_j$, subject to a linear cost function (3.2.3) are*

$[n_i^*]$, *where*

$$n_i^* = \frac{C - c_0}{\sum_{i=1}^{k} \sigma_i \sqrt{c_i}} \frac{\sigma_i}{\sqrt{c_i}}, \qquad i = 1, 2, \ldots, k. \qquad (3.2.4)$$

When the interest is only on estimating the $(k-1)$ differences, $\mu_i - \mu_1$, $i = 2, 3, \ldots, k$; the average variance, A_0, is given by

$$
\begin{aligned}
A_0 &= \frac{1}{k-1} \sum_{i=2}^{k} \mathrm{var}(\bar{X}_{i.} - \bar{X}_{1.}) \\
&= \frac{1}{k-1} \left[\frac{(k-1)\sigma_1^2}{n_1} + \sum_{i=2}^{k} \frac{\sigma_i^2}{n_i} \right].
\end{aligned} \qquad (3.2.5)
$$

Minimizing A_0 subject to a constraint on the linear cost function (3.2.3), one gets the following:

Theorem 3.2.2   *The sample sizes n_i, which minimize the average variance A_0 of the estimators of the $(k-1)$ differences of the parameters, $\mu_i - \mu_1$, subject to a given cost are $[n_i^*] + 1$, where*

$$
\begin{aligned}
n_1^* &= \frac{(C - c_0)}{\{\sigma_1 \sqrt{(k-1)c_1} + \sum_{i=2}^{k} \sigma_i \sqrt{c_i}\}} \sqrt{k-1}\, \frac{\sigma_1}{\sqrt{c_1}}, \\
n_i^* &= \frac{(C - c_0)}{\{\sigma_1 \sqrt{(k-1)c_1} + \sum_{i=2}^{k} \sigma_i \sqrt{c_i}\}} \frac{\sigma_i}{\sqrt{c_i}}, \qquad i = 2, 3, \ldots, k.
\end{aligned} \qquad (3.2.6)
$$

From this theorem one notes that $n_1^*/\sqrt{k-1} = n_2^* = \cdots = n_k^*$. when $\sigma_1^2 = \sigma_2^2 = \cdots = \sigma_k^2$, and $c_1 = c_2 = \cdots = c_k$. In other words, one takes a larger size sample on X_1 compared to the other samples, when population variances are the same and the cost of collecting data from each unit is the same for all populations.

While estimating the $k-1$ differences, $\mu_i - \mu_1$, $i = 2, 3, \ldots, k$; one may minimize the volume of the confidence ellipsoid based on the estimators, $\bar{X}_{i.} - \bar{X}_{1.}$. This is accomplished by minimizing the generalized variance of the estimators $\bar{X}_{i.} - \bar{X}_{1.}$. This generalized variance, V, is given by

$$V = \det\left(\frac{\sigma_1^2}{n_1} J_{k-1, k-1} + D\left(\frac{\sigma_2^2}{n_2}, \frac{\sigma_3^2}{n_3}, \ldots, \frac{\sigma_k^2}{n_k} \right) \right), \qquad (3.2.7)$$

where $J_{k-1,k-1}$ is a $(k-1) \times (k-1)$ matrix of all ones, and $D(.,.,\ldots,.)$ is a diagonal matrix of its arguments. On simplification,

$$V = \frac{\sum_{i=1}^{k} \frac{n_i}{\sigma_i^2}}{\prod_{i=1}^{k} \frac{n_i}{\sigma_i^2}}. \tag{3.2.8}$$

By assuming $\sigma_1^2 = \sigma_2^2 = \cdots = \sigma_k^2$ and $c_1 = c_2 = \cdots = c_k$, one can minimize V subject to the constraint on the cost function to get the following:

Theorem 3.2.3 *When $\sigma_1^2 = \sigma_2^2 = \cdots = \sigma_k^2$ and $c_1 = c_2 = \cdots = c_k$, the generalized variance of $\bar{X}_{i.} - \bar{X}_{1.}$, $i = 2, 3, \ldots, k$ is minimized subject to a given cost, if $n_1 = n_2 = \cdots = n_k$, and the common value, n, is $[n^*] + 1$, where*

$$n^* = \frac{C - c_0}{kc_1}. \tag{3.2.9}$$

3.2.2 Estimation under Normality Assumption

Let $X_i \sim N(\mu_i, \sigma^2)$, $i = 1, 2, \ldots, k$. One may like to estimate all the $k(k-1)/2$ differences, $\mu_i - \mu_j$, simultaneously controlling the absolute error. This specification can be expressed as

$$P(|(\bar{X}_{i.} - \bar{X}_{j.}) - (\mu_i - \mu_j)| \leq d, 1 \leq i < j \leq k) \geq 1 - \alpha, \tag{3.2.10}$$

for given positive d and α. Assuming equal sample sizes $n_1 = n_2 = \cdots = n_k = n$ and σ^2 known, n is determined to satisfy (3.2.10). This condition can be rewritten as

$$P\left(|Z_i - Z_j| \leq d\sqrt{\frac{n}{\sigma^2}}, 1 \leq i < j \leq k\right) \geq 1 - \alpha, \tag{3.2.11}$$

where $Z_i = (\bar{X}_{i.} - \mu_i)/(\sigma/\sqrt{n})$, $i = 1, 2, \ldots, k$ are independent $N(0, 1)$ variables. To satisfy the condition (3.2.11), one needs

$$P\left(Z_{max} - Z_{min} \leq d\sqrt{\frac{n}{\sigma^2}}\right) \geq 1 - \alpha. \tag{3.2.12}$$

Thus,

$$d\sqrt{\frac{n}{\sigma^2}} = q_{k,\infty}^{\alpha},$$ (3.2.13)

where $q_{k,\infty}^{\alpha}$ is the upper α percentile of the range of k independent $N(0, 1)$ variables, tabulated by Harter (1960). Solving (3.2.13), for n, one gets

$$n^* = \left\{\frac{\sigma q_{k,\infty}^{\alpha}}{d}\right\}^2.$$ (3.2.14)

Thus, the following theorem:

Theorem 3.2.4 *The common sample sizes required from each population to control absolute errors specified by Equation (3.2.10) are $[n^*] + 1$, where n^* is given in (3.2.14).*

If the interest centers on estimating the $k - 1$ differences, $\mu_i - \mu_1$, to control the absolute errors, the common sample size n has to be chosen to satisfy the condition

$$P(|(\bar{X}_{i.} - \bar{X}_{1.}) - (\mu_i - \mu_1)| \leq \delta, 2 \leq i \leq k) \geq 1 - \alpha;$$ (3.2.15)

that is,

$$P\left(\max_{2 \leq i \leq k} |(Z_i - Z_1)|/\sqrt{2} \leq \delta\sqrt{\frac{n}{2\sigma^2}}\right) \geq 1 - \alpha.$$ (3.2.16)

This can be accomplished by choosing

$$\delta\sqrt{\frac{n}{2\sigma^2}} = |d|_{k-1,\infty}^{\alpha},$$ (3.2.17)

where $|d|_{k-1,\infty}^{\alpha}$ is the upper 100α percentile of the distribution of the variable

$$\max_{2 \leq i \leq k} \{|Z_i - Z_1|/\sqrt{2}\}$$ (3.2.18)

and are tabulated by Dunnett (1955). Solving (3.2.17) for n,

$$n^* = 2\left\{\frac{\sigma|d|_{k-1,\infty}^{\alpha}}{\delta}\right\}^2.$$ (3.2.19)

Hence the following:

Theorem 3.2.5 *The common sample size n required to control absolute errors specified by (3.2.15) is* $[n^*] + 1$, *when* n^* *is given by (3.2.19).*

When σ is unknown, one may have to use a Stein-type two-stage procedure analogous to the one discussed in Section 2.2 (see Problems 17 and 18).

3.3   Sampling from Normal Distributions

Let $X_i \sim N(\mu_i, \sigma^2)$, for $i = 1, 2, \ldots, k$. Let $X_{i1}, X_{i2}, \ldots, X_{in_i}$ be a random sample on $X_i, i = 1, 2, \ldots, k$. Let the k samples be independent. In this case, one will usually be interested to test the null hypothesis

$$H_0: \mu_1 = \mu_2 = \cdots = \mu_k \tag{3.3.1}$$

against the alternative

$$H_A: \text{not all } \mu_i \text{ are equal}, \tag{3.3.2}$$

assuming σ^2 is unknown.

Let $\bar{X}_{i.}$ and s_i^2 be the ith sample mean and variance and let s_p^2 be the pooled variance. The critical region for the test is

$$F^* = \frac{\sum_{i=1}^{k} n_i (\bar{X}_{i.} - \bar{X}_{..})^2}{(k-1)s_p^2} > F_{1-\alpha}(k-1, n-k), \tag{3.3.3}$$

where $n = \sum_{i=1}^{k} n_i$ and $\bar{X}_{..} = \sum_{i=1}^{k} n_i \bar{X}_{i.}/n$. One likes to determine the sample sizes such that the test has a power of at least $1 - \beta$ when $\mu_i = \mu_i^*$, for $i = 1, 2, \ldots, k$ for given μ_i^*. Under the alternative hypothesis $\mu_i = \mu_i^*$, $i = 1, 2, \ldots, k$, F^* is distributed as a noncentral F variable with $(k-1)$ and $(n-k)$ degrees of freedom, having the noncentrality parameter ϕ given by

$$\phi = \sqrt{\frac{\sum_{i=1}^{k} n_i (\mu_i^* - \tilde{\mu}^*)^2}{k\sigma^2}}, \tag{3.3.4}$$

where

$$\tilde{\mu}^* = \sum_{i=1}^{k} \frac{n_i \mu_i^*}{n}. \qquad (3.3.5)$$

This distribution will be denoted by $F(k-1, n-k; \phi)$. Then the power requirement is

$$P(F^* > F_{1-\alpha}(k-1, n-k)) \geq 1 - \beta, \qquad (3.3.6)$$

where $F^* \sim F(k-1, n-k; \phi)$. The equation can be rewritten as

$$F_{1-\beta}\left(k-1, \sum_{i=1}^{k} n_i - k, \phi\right) \geq F_{1-\alpha}\left(k-1, \sum_{i=1}^{k} n_i - k\right), \quad (3.3.7)$$

where $F_{1-\beta}(.,.,.)$ is $(1-\beta)$ 100 percentile of noncentral distribution, and this equation has to be solved to determine n_is. To get a unique solution to this problem, one may assume $n_i = u_i m$ for prechosen constants u_i based on the cost and other considerations, determine m and then find the required sample sizes. For the special case $n_1 = n_2 = \cdots = n_k = m$, Equation (3.3.7) reduces to

$$F_{1-\beta}(k-1, k(m^*-1); \phi) \geq F_{1-\alpha}(k-1, k(m^*-1)). \quad (3.3.8)$$

This equation can be solved with the help of power curves of F test (see Pearson and Hartley (1951), Odeh and Fox (1975)). Now,

Theorem 3.3.1 *The common sample size m for an α-level test of $H_0: \mu_1 = \mu_2 = \cdots = \mu_k$ is $[m^*] + 1$, where m^* is the smallest positive integer satisfying (3.3.8).*

The main difficulty encompassed in finding sample sizes is the specification of μ_i^* and calculating the noncentrality parameter ϕ. This ϕ contains the nuisance parameter σ^2. Guenther (1977) noted that if μ_i^* are expressed as percentiles of a normal distribution with mean μ, then ϕ will be free from the nuisance parameter σ^2. This approach will be discussed in Example 3.3.1.1. Alternatively, one may desire to detect a difference of at least Δ between two means. Since the power of the F test is an increasing function of ϕ, one can minimize ϕ such that the difference between two means is Δ. This can be achieved by taking two of the μ_i^*s at Δ distance apart and the remaining μ_i^*s to be equal to the average of those two means. The resulting smallest ϕ, $\phi_{\min}$, is

$$\phi_{\min} = \sqrt{\frac{m}{2k}} \frac{\Delta}{\sigma}. \tag{3.3.9}$$

By specifying Δ/σ, $\phi_{\min}$ is a function of m, and the sample sizes are determined from the power curves by iteration. Sample sizes can be determined, however, when α, $1 - \beta$ and Δ/σ are specified, from the tables given by Bowman and Kastenbaum (1975). Example 3.3.1.2 illustrates this approach.

Example 3.3.1.1   Three brands A, B, and C of antacid tablets are to be compared for equality of mean dissolving times using a 0.05 level test. The investigator wants a power of 0.90 to detect the difference when the mean dissolving times of B and C are 40th and 60th percentiles of the distribution of the antacid A. Here $\mu_1^* = \mu_1^*$, $\mu_2^* = \mu_1^* - z_{.4}\sigma = \mu_1^* - 0.2533\sigma$, $\mu_3^* = \mu_1^* + z_{.4}\sigma = \mu_1^* + 0.2533\sigma$. Now $\bar{\mu}^* = \mu_1^*$ and $\phi = 0.2068\sqrt{m}$. The numerator degrees of freedom for the F test is $\nu_1 = 2$ and as m is unknown, initially the denominator degrees of freedom will be taken as $\nu_2 = \infty$. Looking at the Pearson–Hartley (1951) tables of power function, one determines $\phi = 2.05$. Equating the two ϕ values, $m = 98.27$. The initial approximation for m is 99. Then $\nu_2 = 294$. Again ν_2 can be taken as ∞, and thus $m = 99$. In this problem, 99 determinations are needed on each of the three antacids.

Example 3.3.1.2   Again, suppose in the previous antacid problem, the researcher wants to detect a difference of at least 0.506 in σ units with a power of 0.90 using a 0.05 level test. Entering the tables of Bowman and Kastenbaum (1975) with $\Delta/\sigma = 0.506$, $\alpha = 0.05$, and $\beta = 0.1$, one gets $m = 100$. The researcher should take 100 determinations on each of the antacids.

An approximation to the required common sample size, under the setting of Example 3.3.1.2. is (*cf.* Schwertman (1987)) $[m^*] + 1$, where

$$m^* = 2\{\sqrt{\chi^2_{1-\alpha}(k-1)} - (k-2) + z_\beta\}^2 \left(\frac{\sigma}{\Delta}\right)^2 \qquad (3.3.10)$$

Using this formula, the common sample size needed in the ant-acid problem of Example 3.3.1.2 is 97.

By approximating noncentral F distribution with central F distribution with nonintegral degrees of freedom, Guenther (1979) gave a method for determining the sample sizes.

Instead of using an F test, one can use the range test when σ^2 is known, and the studentized range test when σ^2 is unknown. To ensure minimum power over a class of alternatives, when σ^2 is unknown, one can consider a two-stage sampling scheme and base the test on the sample means and s_p^2 (based on first stage samples). The details about this test are given in Hochberg and Lachenbruch (1976).

3.4 Sampling from Bernoulli Distributions

Let X_i be distributed as a Bernoulli variable with θ_i as the parameter, $i = 1, 2, .., k$. Let $X_{i1}, X_{i2}, \ldots, X_{in_i}$ be a random sample on X_i, $i = 1, 2, .., k$. Let the k samples be independent. The null hypothesis of interest in this case is

$$H_0: \theta_1 = \theta_2 = \cdots = \theta_k, \qquad (3.4.1)$$

or equivalently,

$$H_0: \phi_1 = \phi_2 = \cdots = \phi_k, \qquad (3.4.2)$$

where $\phi_i = 2 \arcsin \sqrt{\theta_i}$, $i = 1, 2, \ldots, k$. The null hypothesis (3.4.2) can be rewritten as

$$H_0: \sum_{i=1}^{k} (\phi_i - \bar{\phi}_.)^2 = 0, \qquad (3.4.3)$$

where $\bar{\phi}_.$ is the average of ϕ_is. A commonly used estimator of ϕ_i is

$$\hat{\phi}_i = 2 \arcsin (\sqrt{\bar{X}_{i.}}), \qquad (3.4.4)$$

where $\bar{X}_{i.}$ is the mean of the sample on X_i, $i = 1, 2, \ldots, k$. Noting that $\hat{\phi}_i$ is approximately distributed as $N(\phi_i, 1/n_i)$, the critical region for an approximate α-level test of H_0 given in (3.4.3) is

$$U = \sum_{i=1}^{k} n_i(\hat{\phi}_i - \tilde{\hat{\phi}}_.)^2 > \chi^2_{1-\alpha}(k-1), \qquad (3.4.5)$$

where

$$\tilde{\hat{\phi}}_. = \frac{\sum_{i=1}^{k} n_i \hat{\phi}_i}{\sum_{i=1}^{k} n_i}. \qquad (3.4.6)$$

If one requires a power of at least $1 - \beta$ when $\phi_i = \phi_i^*$, $i = 1, 2, \ldots, k$, the sample sizes n_is will be determined such that

$$P(U > \chi^2_{1-\alpha}(k-1) | \phi_i = \phi_i^*) \geq 1 - \beta. \qquad (3.4.7)$$

When $\phi_i = \phi_i^*$, $i = 1, 2, \ldots, k$, the distribution of the random variable U is a noncentral χ^2 with $k - 1$ degrees of freedom and the noncentrality parameter λ given by

$$\lambda = \sqrt{\sum n_i(\phi_i^* - \tilde{\phi}_.^*)^2}, \qquad (3.4.8)$$

where

$$\tilde{\phi}_.^* = \frac{\sum_{i=1}^{k} n_i \phi_i^*}{\sum_{i=1}^{k} n_i}. \qquad (3.4.9)$$

Denoting a noncentral χ^2 variable with $k - 1$ degrees of freedom and noncentrality parameter λ by $\chi^2(k-1; \lambda)$, (3.4.7) can be rewritten as

$$P(\chi^2(k-1; \lambda) > \chi^2_{1-\alpha}(k-1)) \geq 1 - \beta, \qquad (3.4.10)$$

and consequently,

$$\chi^2_{1-\beta}(k-1; \lambda) \geq \chi^2_{1-\alpha}(k-1), \qquad (3.4.11)$$

where $\chi^2_{1-\beta}(.;.)$ is $100 \, (1 - \beta)$ percentile of a noncentral χ^2 distribution, can be solved for λ using noncentral χ^2 tables, and sample sizes can be determined from that λ value. Thus,

Theorem 3.4.1 *The sample sizes $n_1, n_2, \ldots, n_k$ for an α-level test of H_0: $\theta_1 = \theta_2 = \cdots = \theta_k$ are smallest integer values satisfying (3.4.11).*

In general, specifying individual ϕ_i^* values may not be feasible. One likes to detect a difference of at least Δ among the ϕ_i values. In terms of Δ, the minimum value of λ, when $n_1 = n_2 = \cdots = n_k = m$, is

$$\lambda_{\min} = \frac{\Delta \sqrt{m}}{2}. \qquad (3.4.12)$$

Now m will be determined so that (3.4.11) is satisfied for $\lambda = \lambda_{\min}$. The following example illustrates this method.

Example 3.4.1 Let $k = 3$, $\alpha = 0.05$ and $1 - \beta = 0.90$. From the table prepared by Haynam *et al.* (1973), the λ-value satisfying (3.4.10) is 12.654. Using this λ-value in (3.4.12), the equation that gives the m-value is

$$m = \frac{2(12.654)^2}{\Delta^2}$$

$$= \frac{320.2474}{\Delta^2}.$$

It may be noted that the largest value of Δ is π, so that the minimum common sample size is $320.2474/\pi^2 = 32.4$, that is, 33.

3.5 Sampling from One-Parameter Exponential Distributions

Let the distribution of X_i be the one-parameter (negative) exponential distribution with mean $\theta_i, i = 1, 2, \ldots, k$. Let X_{i1}, $X_{i2}, \ldots, X_{in_i}$ be a random sample on X_i, $i = 1, 2, \ldots, k$. The null hypothesis of interest is

$$H_0: \theta_1 = \theta_2 = \cdots = \theta_k, \qquad (3.5.1)$$

or equivalently,

$$H_0: \phi_1 = \phi_2 = \cdots = \phi_k, \qquad (3.5.2)$$

where $\phi_i = \ln \theta_i$, $i = 1, 2, \ldots, k$. Let $\bar{X}_{i.}$ be the mean of the sample of size n_i on X_i, $i = 1, 2, \ldots, k$. The maximum likelihood estimator of ϕ_i is

$$\hat{\phi}_i = \ln \bar{X}_{i.}, \qquad i = 1, 2, \ldots, k. \tag{3.5.3}$$

The critical region of an approximate α-level test of H_0 given in (3.5.2) is

$$V = \sum_{i=1}^{k} n_i(\hat{\phi}_i - \tilde{\phi}_.)^2 > \chi^2_{1-\alpha}(k - 1), \tag{3.5.4}$$

where

$$\tilde{\phi}_. = \frac{\sum_{i=1}^{k} n_i \hat{\phi}_i}{\sum_{i=1}^{k} n_i}. \tag{3.5.5}$$

If one requires that the power is at least $1 - \beta$ when $\phi_i = \phi_i^*$, $i = 1, 2, \ldots, k$, the sample sizes n_is will be determined so that

$$P(V > \chi^2_{1-\alpha}(k - 1) \mid \phi_i = \phi_i^*) \geq 1 - \beta. \tag{3.5.6}$$

The distribution of the random variable V under the alternative hypothesis can be approximated by a noncentral χ^2 with $k - 1$ degrees of freedom and noncentrality parameter λ given by

$$\lambda = \sqrt{\sum_{i=1}^{k} n_i(\phi_i^* - \tilde{\phi}_.^*)^2} \tag{3.5.7}$$

where

$$\tilde{\phi}_.^* = \frac{\sum_{i=1}^{k} n_i \phi_i^*}{\sum_{i=1}^{k} n_i}. \tag{3.5.8}$$

Using the notation of Section 3.4, the requirement (3.5.6) can be rewritten as

$$P(\chi^2(k - 1; \lambda) > \chi^2_{1-\alpha}(k - 1)) \geq 1 - \beta, \tag{3.5.9}$$

that is,

$$\chi^2_{1-\beta}(k - 1; \lambda) \geq \chi^2_{1-\alpha}(k - 1), \tag{3.5.10}$$

which can be solved for λ, and the sample sizes are then determined. Now,

Theorem 3.5.1   *The sample sizes* $n_1, n_2, \ldots, n_k$ *for an α-level test of H_0: $\theta_1 = \theta_2 = \cdots = \theta_k$ are the smallest integer values satisfying* (3.5.10).

As in Section 3.4, if one likes to detect a difference of at least Δ among the ϕ_i values, the minimum value of λ, when $n_1 = n_2 = \cdots = n_k = m$, is

$$\lambda_{\min} = \Delta \sqrt{\frac{m}{2}}. \tag{3.5.11}$$

Now m will be determined so that (3.5.9) is satisfied for $\lambda = \lambda_{\min}$, using the noncentral χ^2 tables.

Chapter 4   Power and Sample Size

4.1 Introduction

In the earlier chapters, sample sizes for tests of hypotheses were determined by controlling the power of the test at a specified alternative. However, the sample sizes thus determined may be large. Thus, the experimenter wants to examine the power function in order to decide upon a suitable sample size, which is a compromise between the costs and the power of the test.

In this chapter the power functions of the t-test, chi-square test and F-test will be derived. Some important properties of these functions will be mentioned.

When the experimenter is interested in estimating some parameters, and if the sample size determined by the methods discussed earlier is not feasible for experimentation, he can tabulate the bounds on absolute (d^*) or relative (r^*) errors for various sample sizes and can decide the suitable sample size for the experiment.

4.2   Noncentral *t*-Distribution

Let X be a $N(\mu, 1)$ variable and Y be a $\chi^2(\nu)$ variable. Further, let X and V be independent. The distribution of the variable

$$T = \frac{X}{\sqrt{(Y/\nu)}}, \tag{4.2.1}$$

is called a noncentral *t*-distribution with ν degrees of freedom and noncentrality parameter μ. The random variables with this distribution will be denoted by $T(\nu; \mu)$. It may be noted that this distribution reduces to the (central) *t*-distribution when $\mu = 0$. It can be shown that the density function of the noncentral *t*-distribution is

$$f(t; \mu, \nu) = \frac{e^{-\mu^2/2}}{\sqrt{\pi\nu}\,\Gamma\left(\dfrac{\nu}{2}\right)} \sum_{j=0}^{\infty} \left(t\mu\sqrt{\frac{2}{\nu}}\right)^j \frac{\Gamma\left(\dfrac{j+\nu+1}{2}\right)}{j!\left(1+\dfrac{t^2}{\nu}\right)^{(j+\nu+1)/2}}. \tag{4.2.2}$$

The distribution function of T can be written as

$$P(T \le t) = \int_0^{\infty} \Phi\left(t\sqrt{\frac{y}{\nu}} - \mu\right) g(y)\, dy, \tag{4.2.3}$$

where $\Phi(\cdot)$ is the distribution function of the $N(0, 1)$ variable, and $g(\cdot)$ is density function of $\chi^2(\nu)$ variable. From (4.2.3) it is seen that the distribution function of T decreases as μ increases.

Now the power function of the *t*-test considered in the earlier chapters will be expressed in terms of the distribution function of this noncentral *t*-distribution, and the next two sections will be devoted to this discussion.

4.2.1   *Power Function of the One-Sample* t-*Test*

The critical region of an α-level test of $H_0: \mu = \mu_0$ against the alternative $H_A: \mu > \mu_0$, where μ is the mean of a normal distribution, is

$$\bar{X} > \mu_0 + t_\alpha(v)\left(\frac{s}{\sqrt{n}}\right),$$
(4.2.4)

where $\bar{X}$ and s^2 are the mean and variance of a sample of size n and $v = n - 1$. The power function of this test is

$$\pi_1(\mu; v) = P\left(\bar{X} > \mu_0 + t_\alpha(v)\left(\frac{s}{\sqrt{n}}\right)\middle|\mu, \sigma^2\right)$$

$$= P\left(\frac{(\bar{X} - \mu_0)\middle/\sqrt{\dfrac{\sigma^2}{n}}}{\sqrt{\left(\dfrac{s^2}{n}\right)\middle/\left(\dfrac{\sigma^2}{n}\right)}} > t_\alpha(v)\middle| v, \sigma^2\right)$$
(4.2.5)

$$= P(T(v; \Delta) > t_\alpha(v)),$$

where $\Delta = (\mu - \mu_0)\sqrt{n}/\sigma$. For fixed α and v from the monotonicity property of the distribution function of $T(v; \Delta)$ variable, one notes that the power function of this *t*-test increases with Δ. It may be noted that the power function depends on the nuisance parameter σ^2.

Let $\mu_2 > \mu_1$ be two values of μ in the set $\{\mu | \mu > \mu_0\}$ specified by the alternative hypothesis. Then

$$\pi_1(\mu_2; v) > \pi_1(\mu_1; v) > \pi_1(\mu_0; v) = \alpha.$$
(4.2.6)

In other words, if the power at $\mu = \mu_1$ is set at $1 - \beta$, then the power at $\mu = \mu_2$ is greater than $1 - \beta$. Thus it is sufficient to control the power at some specified alternative μ_1, where μ_1 is the smallest value that is judged to be so different from μ_0 by the investigator. This observation is the motivating factor in determining the sample sizes by controlling the power of an α-level test at a specified alternative hypothesis.

Due to budgetary and other constraints, it may not be feasible for the experimenter to use the sample size determined. In such cases, one may like to examine the power of the test for other sample sizes. Such an examination will enable the experimenter to decide upon the appropriate sample size making a compromise between the power at the specified alternative and other constraints. For the chosen sample size, one may like to know what type of differences in the mean can be detected at the prechosen power $1 - \beta$. This difference can be found by solving $\pi_1(\mu; \nu) = 1 - \beta$ for μ.

Now consider the problem of testing $H_0: \mu = \mu_0$ against the two-sided alternative $H_A: \mu \neq \mu_0$. The critical region of the usual α-level test is

$$|\bar{X} - \mu_0| > t_{\alpha/2}(\nu)\left(\frac{s}{\sqrt{n}}\right). \tag{4.2.7}$$

The power function of this test using t for $t_{\alpha/2}(\nu)$ is

$$\pi_2(\mu; \nu) = P(|T(\nu; \Delta)| > t)$$
$$= P(T(\nu; \Delta) > t) + P(T(\nu; \Delta) < -t)$$
$$= 1 - \int_0^\infty \Phi\left(t\sqrt{\frac{y}{\nu}} - \Delta\right)g(y)\,dy$$
$$+ \int_0^\infty \Phi\left(-t\sqrt{\frac{y}{\nu}} - \Delta\right)g(y)\,dy$$
$$= 2 - \int_0^\infty \left\{\Phi\left(t\sqrt{\frac{y}{\nu}} - \Delta\right) + \Phi\left(t\sqrt{\frac{y}{\nu}} + \Delta\right)\right\}g(y)\,dy, \tag{4.2.8}$$

from (4.2.3). It can be seen that the power at Δ is same as the power at $-\Delta$. By examining the derivative of (4.2.8) with respect to Δ, it can be seen that the power function increases with $|\Delta|$, and the earlier discussion for the one-sided test is valid here also.

4.2.2 Power Function of the Two-Sample t-Test

The critical region of the usual α-level test of $H_0: \mu_1 = \mu_2$ against the one-sided alternative $H_A: \mu_1 > \mu_2$, where μ_1 and μ_2 are the means of two normal distributions with common unknown variance σ^2, is

$$\bar{X}_1 - \bar{X}_2 > t_\alpha(\nu)\sqrt{s_p^2\left(\frac{1}{n_1} + \frac{1}{n_2}\right)}, \tag{4.2.9}$$

ν being $n_1 + n_2 - 2$. The power function of this test depends on

$$(\mu_1 - \mu_2) \Big/ \Big\{ \sigma \sqrt{\frac{1}{n_1} + \frac{1}{n_2}} \Big\} = \delta \Big/ \Big\{ \sigma \sqrt{\frac{1}{n_1} + \frac{1}{n_2}} \Big\} = \Delta \text{ and is}$$

$$\pi_2(\delta; \nu) = P(T(\nu; \Delta) > t_\alpha(\nu)). \tag{4.2.10}$$

This is an increasing function of δ and also depends on σ^2. Further,

$$\pi_2(\delta_2; \nu) > \pi_2(\delta_1; \nu) > \pi_2(0; \nu) = \alpha, \qquad \text{for } \delta_2 > \delta_1 > 0. \tag{4.2.11}$$

Thus it is sufficient to control the power at $\delta = \delta_1$, where δ_1 is a value so that μ_1 and $\mu_1 + \delta_1$ are considered so different by the investigator.

The power function for the two-sided t-test can be easily determined, and the monotone properties can be derived easily.

4.3   Noncentral χ^2-Distribution

Let $X_1, X_2, \ldots, X_k$ be k independent random variables and $X_i \sim N(\mu_i, 1)$, $i = 1, 2, \ldots, k$. The distribution of the random variable

$$W = \sum_{i=1}^{k} X_i^2 \tag{4.3.1}$$

is called a noncentral chi-square distribution with k degrees of freedom and noncentrality parameter $\tau = \sqrt{\Sigma_{i=1}^{k} \mu_i^2}$. The random variable with this distribution will be denoted by $\chi^2(k; \tau)$. When $\mu_i = 0$, $i = 1, 2, \ldots, k$, a noncentral $\chi^2(k; \tau)$ variable becomes a central $\chi^2(k)$ variable. It can be shown that the density function of a noncentral $\chi^2(k; \tau)$ variable is

$$f(w; k, \tau) = \frac{e^{-(w + \tau^2)/2}}{2^{k/2}} \sum_{j=0}^{\infty} \frac{w^{(k/2) + j - 1} \tau^{2j}}{\Gamma\left(\frac{k}{2} + j\right) 2^{2j} j!}, \tag{4.3.2}$$

and that the distribution function of W is a decreasing function of τ ([see Alam and Rizvi (1967)]).

Consider the problem of testing the equalities of the means of k independent $NE(0, \theta_i)$ variables. The power function of the test suggested in Section 3.5 is

$$\pi(\lambda; k-1) = P(\chi^2(k-1; \lambda) > \chi^2_{1-\alpha}(k-1)), \qquad (4.3.3)$$

where λ is defined by (3.5.7). This is an increasing function of λ. The power depends on the specification θ_i^*, of the means in the alternative hypothesis, only through λ. It may be noted that the λ-value for two sets of means θ_i^* and $r\theta_i^*$, $i = 1, 2, \ldots, k$ for a positive r are same. Hence, the power at these two sets of means are same. In other words, the power depends on the ratios such as θ_j/θ_1, $j = 2, 3, \ldots, k$. Further, when $n_1 = n_2 = \cdots = n_k = (m, \text{ say})$,

$$\lambda^2 = m \sum_{i=1}^{k} (\phi_i^* - \bar{\phi}^*)^2 \geqslant \frac{m\{\phi_{max}^* - \phi_{min}^*\}^2}{2}, \qquad (4.3.4)$$

where $\phi_i^* = \ln \theta_i^*$ and $\bar{\phi}^* = \Sigma_{i=1}^{k} \phi_i^*/k$. Now,

$$\lambda^2_{min} = m\left\{\frac{\phi_{max}^* - \phi_{min}^*}{2}\right\}^2 = \frac{m}{2}\left\{\ln\left(\frac{\theta_{max}^*}{\theta_{min}^*}\right)\right\}^2. \qquad (4.3.5)$$

It may be easy for the experimenter to specify the largest ratio of the means he wanted to detect in the experiment and in such cases λ_{min} can be calculated from (4.3.5), and the power at λ_{min} is controlled at $1 - \beta$ to determine the common sample size.

When it is not feasible to run the experiment with the determined sample sizes, for other values of the sample sizes, the λ-values can be determined from which the largest ratio of the means that could be detected by the test can be obtained. From such an analysis, the experimenter can decide a compromise sample size.

4.4   Noncentral F-Distribution

Let W and V be independent random variables such that $W \sim \chi^2(\nu_1; \tau)$ and $V \sim \chi^2(\nu_2)$. The distribution of the variable

$$U = \frac{(W/\nu_1)}{(V/\nu_2)}, \qquad (4.4.1)$$

is called the noncentral F-distribution with ν_1 and ν_2 degrees of freedom and the noncentrality parameter τ. The random variable with this distribution will be denoted by $F(\nu_1, \nu_2; \tau)$. When $\tau = 0$, the noncentral F-distribution reduces to the central F-distribution. The density function of a noncentral F-distribution is known to be

$$f(u; \nu_1, \nu_2, \tau) = \frac{\nu_1 e^{-\tau^2/2}}{\nu_2 \Gamma(\nu_2/2)} \sum_{j=0}^{\infty} \frac{(\tau^2/2)^j (\nu_1 u/\nu_2)^{(\nu_1/2)+j-1}}{j! \Gamma((\nu_1/2)+j)}$$

$$\times \frac{\Gamma((\nu_1 + \nu_2)/2 + j)}{(1 + (\nu_1 u/\nu_2))^{((\nu_1+\nu_2)/2+j)}}. \qquad (4.4.2)$$

It can be shown that the distribution function of U is a decreasing function of τ.

Consider the problem of testing the equality of the means of k independent $N(\mu_i, \sigma^2)$ variables (See Section 3.3). The power of the F test is

$$\pi(\phi; k-1, n-k) = P(F(k-1, n-k; \phi) > F_{1-\alpha}(k-1, n-k)), \qquad (4.4.3)$$

where ϕ is given by (3.3.4). This power is an increasing function of ϕ. At any specified alternative $\mu_1^*, \mu_2^*, \ldots, \mu_k^*$, the power depends on them only through ϕ. When $n_1 = n_2 = \cdots = n_k \ (= m,$ say) then

$$\phi^2 = \frac{m}{k\sigma^2} \sum_{j=1}^{k} (\mu_i^* - \bar{\mu}^*) \geq \frac{m}{2k\sigma^2} \{\mu_{\max}^* - \mu_{\min}^*\}^2, \qquad (4.4.4)$$

where $\bar{\mu}^* = \sum_{i=1}^{k} \mu_i^*/k$. If one specifies the largest difference in the means to be detected by the test, $\phi_{\min}$ can be calculated and the power at $\phi_{\min}$ is controlled at $1 - \beta$ to determine the common sample size.

If the determined sample size is not convenient to be used in the study, the experimenter can find ϕ for other choices of sample sizes and determine the largest differences that can be detected by the test. From such an analysis, the appropriate sample size can be decided by the experimenter.

Chapter 5 — Samples from Finite Populations

5.1 Introduction

While sampling from finite populations, the investigator is interested mainly in estimating the population parameters such as mean, total and proportion. Usually the tests of hypothesis are not of much interest here. The methodology discussed in Chapters 1, 2 and 3 in estimating the parameters is not applicable in this situation because the exact distribution of the estimators are generally unknown. One notes that controlling absolute error or relative error in estimation can be achieved by controlling the variance or coefficient of variation of the estimator (See Problem one of the Problems and Supplements section of this monograph). In this chapter, the variance of the estimator will be controlled, and the required sample sizes are determined.

5.2 Simple Random Sampling

Consider a population of size N from which a simple random sample of size n is drawn, without replacement. Let $\bar{x}$ be the sample mean and let $\bar{X}$ and S^2 be the population mean and variance. When the response variable is dichotomous taking the values 0 or 1, $\bar{x}$ and $\bar{X}$ will be denoted by p and P, respectively. In this case, $S^2 = NPQ/(N-1)$, where $Q = 1 - P$.

It is known that

$$\text{Var}(\bar{x}) = \left(\frac{N-n}{Nn} \right) S^2. \tag{5.2.1}$$

By imposing the restraint,

$$\text{Var}(\bar{x}) \leq V^*, \tag{5.2.2}$$

for a prechosen V^*, n is determined to satisfy this inequality. The required n is

$$n \geq \frac{NS^2}{S^2 + NV^*} = n^*, \tag{5.2.3}$$

and the following is established.

Theorem 5.2.1 *The sample size n, required with the specification (5.2.2) while random sampling without replacement from a finite population of size N is $[n^*] + 1$, where n^* is given in (5.2.3).*

Prior information on S^2 is needed to determine the sample size using Theorem 5.2.1. Such information can be obtained from (a) previous studies, or (b) pilot surveys. If the prior information on S^2 is of doubtful validity, one may use a two-stage procedure as described next:

(i) Take an initial sample of size $n_1(\geq 2)$. Calculate s_1^2, the variance of this sample. s_1^2 is an unbiased estimator of S^2.

(ii) Calculate

$$n_2 = \frac{s_1^2}{V^*} \left(1 + \frac{2}{n_1} \right),$$

$$n = \max \left\{ n_1, \left[n_2 \frac{1}{\left(\frac{n_2}{N} \right) + 1} \right] + 1 \right\}. \tag{5.2.4}$$

(iii) Take $n - n_1$ additional observations, if necessary. The sample mean $\bar{x}$ based on all n observations is the required estimator of $\bar{X}$.

For details of this procedure, see Cochran (1977) and Cox (1952).

Example 5.2.1.1   In a survey aimed at estimating the average number of working persons in a household in a community of $N = 10{,}000$ households, it is assumed that $S^2 = 36$ based on a pilot study. The required sample size with $V^* = (0.25)^2$ is

$$\left[\frac{10000(36)}{36 + (10000)(0.25)^2} \right] + 1 = 545.$$

When the response variable is dichotomous taking values 0 or 1, the expression for n^* of Equation (5.2.3) becomes

$$n_p^* = \frac{N}{1 + \dfrac{(N-1)V^*}{PQ}}. \tag{5.2.5}$$

Since P is usually unknown, and is somewhat difficult to guess as the parameter being estimated is P itself, a conservative approach is to take $P = \frac{1}{2}$. Thus,

Theorem 5.2.2   *The sample size n required to estimate a population proportion p such that* $\mathrm{Var}\,(p) \leqslant V^*$ *is* $[n_{0.5}^*] + 1$, *where n_p^* is given by* (5.2.5).

Example 5.2.2.1   Consider a community with $N = 10{,}000$ registered voters. To estimate the proportion of votes cast in favor of candidate A, with $V^* = (0.025)^2$, the required sample size at an exit poll is

$$[10{,}000/\{1 + (9{,}999)(0.000625)/((0.5)(0.5))\}] + 1 = 385.$$

In the following section, determination of sample size in the context of acceptance sampling will be discussed.

5.2.1   Acceptance Sampling

The population of interest consists of N manufactured items of which $D = NP$ are defective. Thus, the remaining $N - D = NQ$ items are good. Usually D is not known. On the basis of a sample of size n, one likes to test the null hypothesis H_0: $D = D_0$ against the alternative H_A: $D > D_0$. An α-level test of this H_0 has critical region

$$Y > c, \qquad\qquad (5.2.6)$$

where Y is the number of defective items in the sample, and a positive integer c is determined such that

$$\sum_{y=c+1}^{\min(D_0,n)} \frac{\binom{D_0}{y}\binom{N - D_0}{n - y}}{\binom{N}{n}} \leq \alpha. \qquad\qquad (5.2.7)$$

If one desires to have a power of no less than $1 - \beta$ when $D = D_1$ ($>D_0$), the requirement is

$$\sum_{y=c+1}^{\min(D_1,n)} \frac{\binom{D_1}{y}\binom{N - D_1}{n - y}}{\binom{N}{n}} \geq 1 - \beta. \qquad\qquad (5.2.8)$$

Equations (5.2.7) and (5.2.8) should be solved for n. The probabilities used in (5.2.7) and (5.2.8) are based on hypergeometric distribution and are tabulated by Lieberman and Owen (1961). An approximate solution for this problem was considered by Guenther (1977).

One can approximate the distribution of Y by $N(nP, nPQ)$ and get the following result.

Theorem 5.2.3   *An approximation to the sample size n required for an α-level test of H_0: $D = D_0$ against the one-sided alternative H_A: $D > D_0$ to give a power of at least $1 - \beta$ when $D = D_1$ ($>D_0$)*

is $[n^*] + 1$, *where* n^* *is given by*

$$n^* = \left\{ \frac{z_\alpha \sqrt{D_0(N - D_0)} + z_\beta \sqrt{D_1(N - D_1)}}{(D_1 - D_0)} \right\}^2. \qquad (5.2.9)$$

Example 5.2.3.1   A purchaser bought a lot of 100 items. He can accept the lot if it has at most five defectives, and he likes to definitely reject the lot with a high probability if it has 20 or more defectives. This acceptance problem can be formulated as testing $H_0: D = 5$, against $H_A: D > 5$. Suppose the purchaser wants to use a 0.05 level test with a power of 0.90 at $D = 20$. Thus, $N = 100$, $D_0 = 5$, $D_1 = 20$, $\alpha = 0.05$ and $\beta = 0.10$, and from Equation (5.2.9),

$$n^* = \left\{ \frac{(1.645)(\sqrt{5(95)}) + (1.282)(\sqrt{20(80)})}{(20 - 5)} \right\}^2$$

$$= 33.74.$$

The sample size needed for this problem is 34. To implement this plan, one notes that the lot is rejected, if the number of defectives in the sample of 34 items is more than four, a number derived from Equation (5.2.7).

5.3   Stratified Random Sampling

Consider a population that is stratified into L strata. The hth strata is of size N_h and has population variances S_h^2, for $h = 1$, $2, \ldots, L$. If the experimenter takes a stratified random sample of size n, it is necessary to decide the allocation of this total size n among the strata. Let n_h be the sample size allocated to strata h, $h = 1, 2, \ldots, L$.

Let $\bar{X}$ be the population mean. Let $\bar{x}_h$ be the sample mean from the hth stratum. It is then known that

$$\hat{\bar{X}} = \sum_{h=1}^{L} \frac{N_h \bar{x}_h}{N}, \qquad (5.3.1)$$

where $N = \Sigma_{h=1}^{L} N_h$ is the total population size, is an unbiased estimator of $\bar{X}$. Further

$$\text{Var}(\hat{\bar{X}}) = \frac{1}{N^2} \sum_{h=1}^{L} N_h(N_h - n_h) \frac{S_h^2}{n_h}. \qquad (5.3.2)$$

The allocation of the total sample size into the strata is called proportional allocation if

$$n_h = \frac{nN_h}{N}, \qquad (5.3.3)$$

and in that case

$$\text{Var}_{\text{Prop}}(\hat{\bar{X}}) = \frac{N-n}{N^2 n} \sum_{h=1}^{L} N_h S_h^2. \qquad (5.3.4)$$

If one wants to control this variance by imposing the condition,

$$\text{Var}_{\text{Prop}}(\hat{\bar{X}}) \leq V^*, \qquad (5.3.5)$$

then the required total sample size n is $[n^*] + 1$, where n^* is given by

$$n^* = N \left\{ \frac{\Sigma_{h=1}^{L} N_h S_h^2}{V^* N^2 + \Sigma_{h=1}^{L} N_h S_h^2} \right\}. \qquad (5.3.6)$$

Thus the following:

Theorem 5.3.1 *Under proportional allocation, the total sample size needed to satisfy the condition (5.3.5) is $[n^*]+1$, where n^* is given by the Equation (5.3.6). Further, the sample size for the hth stratum is $[n^* n_h / N] + 1$, for $h = 1, 2, \ldots, L$.*

Instead of using the proportional allocation, one may like to use the optimum allocation such that $\text{Var}(\hat{\bar{X}})$ is minimum subject to a cost constraint using a linear cost function

$$c_0 + \sum_{h=1}^{L} c_h n_h = C; \qquad (5.3.7)$$

the required n_h values can be derived from Cauchy-Schwartz inequality after approximating $N_h - n_h$ by N_h in (5.3.2) and are given below:

Theorem 5.3.2 *The n_h-values that minimize* $\text{Var}(\hat{\bar{X}})$ *subject to the cost constraint (5.3.7) are* $[n_h^*] + 1$, *for* $h = 1, 2, \ldots, L$, *where* n_h^* *are*

$$n_h^* = \frac{(C - c_0) N_h S_h / \sqrt{c_h}}{\sum_{h=1}^{L} N_h S_h \sqrt{c_h}}, \qquad h = 1, 2, \ldots, L \qquad (5.3.8)$$

Similar results can be easily obtained when one is estimating the population proportions.

5.4 Double Sampling for Stratification

Consider a stratified population where there are L strata, and the total population size N is divided into these strata of sizes N_1, $N_2, \ldots, N_L$. However, the investigator has no knowledge of these strata sizes. In that case, initially a simple random sample of size n' can be taken and based on this sample, the investigator can estimate the strata sizes by $N n_h' / n'$, $h = 1, 2, \ldots, L$, where n_h' is the number of units in this first sample falling in stratum h. Now a second phase sample of $n_h = v_h n_h'$ units are randomly drawn from the n_h' units selected in the first phase. If $\bar{x}_h$ is the mean of the second phase h^{th} stratum sample, then the population mean $\bar{X}$ is estimated by

$$\hat{\bar{X}} = \sum_{h=1}^{L} \frac{n_h'}{n'} \bar{x}_h, \qquad (5.4.1)$$

and an approximate $\text{Var}(\hat{\bar{X}})$ is given by

$$\text{Var}(\hat{\bar{X}}) = S^2 \left(\frac{1}{n'} - \frac{1}{N} \right) + \sum_{h=1}^{L} \frac{N_h S_h^2}{N n'} \left(\frac{1}{v_h} - 1 \right), \qquad (5.4.2)$$

where S^2 is the population variance of all N units, and S_h^2 is the population variance of the hth stratum, $h = 1, 2, \ldots, L$ (*cf.* Cochran (1977), p. 328). Using a cost function

$$C = c_0 + c' n' + \sum_{h=1}^{L} c_h n_h, \qquad (5.4.3)$$

one notes by Cauchy–Schwartz inequality that $\mathrm{Var}(\hat{\bar{X}})$ given in (5.4.2) is minimized for a given expected cost

$$E(C) = c_0 + c'n' + n' \sum_{h=1}^{L} c_h \nu_h \frac{N_h}{N} = C^*, \qquad (5.4.4)$$

when

$$\nu_h = S_h \left\{ c' \Big/ \left(c_h \left(S^2 - \sum_{h=1}^{L} \frac{N_h S_h^2}{N} \right) \right) \right\}^{1/2}. \qquad (5.4.5)$$

The first phase sample size n' can now be found by substitution for ν_h from (5.4.5) in (5.4.4) and is $[n'^*] + 1$, where

$$n'^* = \{C^* - c_0\} \Big/ \left\{ c' + \sum_{h=1}^{L} c_h \nu_h N_h/N \right\}. \qquad (5.4.6)$$

Thus the following:

Theorem 5.4.1 *The approximate size of the initial simple random sample in a double sampling for stratification problem is $[n'^*] + 1$, where n'^* is given by (5.4.6). If n'_h of the initial phase units belong to stratum h, the number of units selected in the second phase from stratum h is $[n'_h \nu_h] + 1$, where ν_h is given by (5.4.5).*

   The double sampling (or two-phase sampling) discussed in this section is different from the two-stage procedures discussed in Section 5.2. In a two-stage procedure, the second stage sample is taken from the remaining population, while in double sampling, the second phase sample is a subset of the first phase sample.

5.5 Two-Stage Cluster Sampling

Let the population consist of N clusters (also known as primary sampling units) and let M_i be the number of subunits (also known as second-stage units) in the ith cluster, $i = 1, 2, \ldots, N$. Let Y_{ij} be the response on the jth second-stage unit within the ith cluster. Let $Y_i = \sum_{j=1}^{M_i} Y_{ij}$, $\bar{Y}_i = Y_i/M_i$, $Y = \sum_{i=1}^{N} Y_i$, $\bar{Y} = Y/N$, $S_{wi}^2 = \sum_{j=1}^{M_i} (Y_{ij} - \bar{Y}_i)^2/(M_i - 1)$. Here Y_i is the ith cluster total response, Y is the total response of all clusters, $\bar{Y}_i$ is the mean of second-stage unit responses in the ith cluster, and $\bar{Y}$ is the mean cluster response. Further, S_{wi}^2 is the variance of the second-stage unit responses within the ith cluster.

A sample of n clusters with probabilities proportional to z_i are selected *with replacement*. Within the ith selected cluster, a sample of m_i second-stage units will be selected by using a random sampling scheme without replacement. Let $\bar{y}_i$ be the sample mean of the second stage unit responses in the ith sampled cluster, $i = 1, 2, \ldots, n$. Then it is well known that

$$\hat{Y} = \sum_{i=1}^{n} \frac{M_i \bar{y}_i}{nz_i} \qquad (5.5.1)$$

is an unbiased estimator of Y, the population total. The variance of $\hat{Y}$ is given by

$$\text{Var}(\hat{Y}) = \frac{1}{n} \left[V_z + \sum_{i=1}^{N} \frac{1}{z_i} M_i^2 \left(\frac{1}{m_i} - \frac{1}{M_i} \right) S_{wi}^2 \right], \qquad (5.5.2)$$

where

$$V_z = \sum_{i=1}^{N} z_i \left(\frac{Y_i}{z_i} - Y \right)^2, \qquad (5.5.3)$$

(*cf.* Cochran (1977), p. 307). Let c_0 be the fixed cost, c_1 be the variable cost to sample the cluster, c_2 be the variable cost to list second-stage units in a selected cluster, and c_3 be the variable cost to sample each second-stage unit. Then the total cost of the survey is $C = c_0 + c_1 n + c_2 \sum_{i=1}^{n} M_i + c_3 \sum_{i=1}^{n} m_i$, and its expectation is given by

$$E(C) = c_0 + c_1 n + c_2 n \sum_{i=1}^{N} z_i M_i + c_3 n \sum_{i=1}^{N} z_i m_i. \qquad (5.5.4)$$

Now for a fixed expected cost, C^*, of the survey, by Cauchy–Schwartz inequality, $\text{Var}(\hat{Y})$ is minimized, when

$$\sqrt{n \left(c_1 + c_2 \sum_{i=1}^{N} z_i M_i \right)} \propto \sqrt{\frac{1}{n} \left(V_z - \sum_{i=1}^{N} \frac{M_i S_{wi}^2}{z_i} \right)},$$

$$\qquad (5.5.5)$$

$$\sqrt{c_3 n z_i m_i} \propto \sqrt{\frac{1}{n} \frac{M_i^2}{z_i m_i} S_{wi}^2}.$$

From (5.5.5) it follows that

$$m_i^* = \frac{M_i S_{wi}}{z_i \sqrt{V_z - \sum_{i=1}^{N} \frac{M_i S_{wi}^2}{z_i}}} \sqrt{\frac{c_1 + c_2 \sum_{i=1}^{N} z_i M_i}{c_3}}. \qquad (5.5.6)$$

Substituting the m_i^* values in the right-hand side of (5.5.4) and equating to C^*, n^* can be obtained. Thus the following:

Theorem 5.5.1 *In a two-stage cluster sampling, where the clusters are selected with replacement and with probabilities proportional to z_i, the variance given in (5.5.2) of the estimator $\hat{Y}$ is minimized by taking $[m_i^*] + 1$ second stage units from the ith selected cluster, where m_i^* is given by (5.5.6). The number of clusters to be selected is $[n^*] + 1$, where*

$$n^* = \frac{C^* - c_0}{c_1 + c_2 \sum_{i=1}^{N} z_i M_i + c_3 \sum_{i=1}^{N} z_i m_i^*}, \qquad (5.5.7)$$

C^* being the specified total expected cost.

   A special case of this setting when the values of the relevant design parameters, the variance ratio and cost ratio are not known exactly but can be assumed to be within some specific range, was considered by Sadooghi-Alvandi (1986).

Chapter 6   Ranking and Selection

6.1   Introduction

Suppose there are k populations $\pi_1, \pi_2, \ldots, \pi_k$. Let θ_i be an unknown parameter associated with the population π_i, $1, 2, \ldots, k$. The ordered values of θs are

$$\theta_{[1]} \leq \theta_{[2]} \leq \cdots \leq \theta_{[k]}, \qquad (6.1.1)$$

and it is unknown to which population $\theta_{[i]}$ belongs, $i = 1$, $2, \ldots, k$. The investigator wants to identify the population that corresponds to $\theta_{[k]}$ (or $\theta_{[1]}$). This population of interest will be called the best population. Now one takes independent samples from these populations of common size n and uses the appropriate estimators of θs, to select the best population. The problem is to determine the common sample size so that the best population is selected with high probability.

In the following sections, results are obtained for normal, Bernoulli and exponential populations.

6.2 Selecting the Best Normal Distribution

The distribution of the observations from the population π_i is $N(\mu_i, \sigma_i^2)$, $i = 1, 2, \ldots, k$. Assuming $\sigma_1^2 = \sigma_2^2 = \cdots = \sigma_k^2$, the problem of selecting the population with largest mean will be considered first. Next, the problem of selecting the population with smallest variance will be considered.

6.2.1 Selecting the Normal Distribution with the Largest Mean When the Common Variance σ^2 Is Known

Let $\bar{X}_i$ be the mean of the sample of size n from π_i, $i = 1, 2, \ldots, k$. The ordered sample means are

$$\bar{X}_{[1]} \leq \bar{X}_{[2]} \leq \cdots \leq \bar{X}_{[k]}. \qquad (6.2.1)$$

The selection rule is to choose the population corresponding to the largest sample mean $\bar{X}_{[k]}$. When there are ties for largest mean, they should be broken at random. One likes to choose the common sample size n such that the probability of a correct selection (cs) is at least P^*, whenever $\mu_{[k]} - \mu_{[k-1]} \geq \delta^*$, for specified positive δ^*. One can correctly choose the best population by random selection with a probability of $1/k$. In order to do better than the random selection in developing a procedure, it is assumed that $P^* > 1/k$. If $\bar{X}_{(i)}$ is the mean of the sample from the population with the parameter $\mu_{[i]}$, $i = 1, 2, \ldots, k$, then

$$P(cs) = P(\bar{X}_{(k)} > \bar{X}_{(i)}, i = 1, 2, \ldots, k - 1). \qquad (6.2.2)$$

Since this probability depends on $\mu_{[1]}, \mu_{[2]}, \ldots, \mu_{[k]}$, which are unknown, the requirement on n is satisfied if

$$\min P(cs) \geq P^*, \qquad (6.2.3)$$

where the minimum is over all parameter values such that $\mu_{[k]} - \mu_{[k-1]} \geq \delta^*$.

The probability of (6.2.2) can be rewritten as

$$P(cs) = P\left(\frac{(\bar{X}_{(k)} - \mu_{[k]})}{\frac{\sigma}{\sqrt{n}}} > \frac{(\bar{X}_{(i)} - \mu_{[i]}) + (\mu_{[i]} - \mu_{[k]})}{\frac{\sigma}{\sqrt{n}}}, \right.$$

$$\left. i = 1, 2, \ldots, k - 1 \right)$$

$$= P\left(Z_k > Z_i - \frac{(\mu_{[k]} - \mu_{[i]})}{\frac{\sigma}{\sqrt{n}}}, \qquad i = 1, 2, \ldots, k - 1 \right)$$

$$\geq P\left(Z_k > Z_i - \left(\frac{\delta^* \sqrt{n}}{\sigma}\right), i = 1, 2, \ldots, k - 1 \right), \qquad (6.2.4)$$

where the Z_is are independent $N(0, 1)$ variables. Thus the requirement (6.2.3) will be met by choosing n so as to satisfy

$$\int_{-\infty}^{\infty} \Phi^{k-1}\left(x + \left(\frac{\delta^* \sqrt{n}}{\sigma}\right)\right) \phi(x)\, dx \geq P^*, \qquad (6.2.5)$$

where $\phi(\cdot)$ is the density function of the $N(0, 1)$ variable. Hence n must be chosen so that

$$n \geq \left(\frac{\lambda \sigma}{\delta^*}\right)^2 = n^*, \qquad (6.2.6)$$

where λ is the solution of

$$\int_{-\infty}^{\infty} \Phi^{k-1}(x + \lambda) \phi(x)\, dx = P^*. \qquad (6.2.7)$$

These λ-values have been tabulated by Bechhofer (1954) and Desu and Sobel (1968) for various k and P^* values. Thus, the following is established.

Theorem 6.2.1   *The common sample size n required for selecting the normal distribution with the largest mean so that the P(cs) $\geq$ P**

whenever $\mu_{[k]} - \mu_{[k-1]} \geq \delta^*$, *is* $[n^*] + 1$, *where* n^* *is given by*
(6.2.6).

Example 6.2.1.1 Let $k = 3$, $P^* = 0.99$, $\sigma = 50$ and $\delta^* = 25$.
From tables the λ-value is 3.62; so

$$n^* = \left\{ (3.62) \frac{50}{25} \right\}^2 = 52.42.$$

Thus, the required sample size is 53.

6.2.2 Selecting the Normal Distribution with the Largest Mean When the Common Variance σ^2 Is Unknown

In this case one may use a two-stage procedure as described next.

(i) Take independent first stage samples of same arbitrary size n_1 from each one of the k populations. Let s_p^2 be the pooled estimator of the common variance.

(ii) Let $n = \max \{n_1, [2s_p^2(h/\delta^*)^2] + 1\}$, where h is from the table prepared by Bechhofer *et al.* (1954).

(iii) Take $n - n_1$ additional observations, if necessary, from each one of the k populations.

(iv) Let $\bar{X}_i$ be the mean of all n observations from π_i, $i = 1, 2, \ldots, k$. Select the population corresponding to maximum $\bar{X}_i$.

6.2.3 Selecting the Normal Distribution with the Smallest Variance

Let $X_{i1}, X_{i2}, \ldots, X_{in}$ be a random sample of size n from π_i, $i = 1, 2, \ldots, k$. A commonly used estimator of σ_i^2 is

$$
\begin{aligned}
\hat{\sigma}_i^2 &= \sum_{j=1}^{n} \frac{(X_{ij} - \bar{X}_{i.})^2}{(n-1)}, && \text{if } \mu_i \text{ is unknown}, \\
&= \sum_{j=1}^{n} \frac{(X_{ij} - \mu_i)^2}{n}, && \text{if } \mu_i \text{ is known}.
\end{aligned}
\tag{6.2.8}
$$

In this context it will be assumed that all the k means are either known or unknown. The ordered values of $\hat{\sigma}_i^2$ are

$$\hat{\sigma}_{[1]}^2 \leqslant \hat{\sigma}_{[2]}^2 \leqslant \cdots \leqslant \hat{\sigma}_{[k]}^2. \tag{6.2.9}$$

The selection rule is to choose the population corresponding to $\hat{\sigma}_{[1]}^2$, the smallest estimated variance. As before, one wants to choose the common sample size n such that the probability of a correct selection is at least P^*, whenever $(\sigma_{[2]}^2/\sigma_{[1]}^2) \geqslant r^*$, where $\sigma_{[1]}^2 \leqslant \sigma_{[2]}^2 \leqslant \cdots \leqslant \sigma_{[k]}^2$ are the ordered values of the population variances, and $P^*(>1/k)$ and $r^*(>1)$ are specified positive constants. If $\hat{\sigma}_{(i)}^2$ is the variance estimator from the population with the parameter $\sigma_{[i]}^2$, $i = 1, 2, \ldots, k$, then

$$P(cs) = P(\hat{\sigma}_{(1)}^2 < \hat{\sigma}_{(i)}^2, i = 2, 3, \ldots, k). \tag{6.2.10}$$

As in Section 6.2.1, the requirement on n is satisfied if

$$\min P(cs) \geqslant P^*, \tag{6.2.11}$$

where the minimum is over all parameter values $\sigma_{[2]}^2 \geqslant r^* \sigma_{[1]}^2$. The probability of (6.2.10) is

$$P(cs) = P\left(\frac{f\hat{\sigma}_{(1)}^2}{\sigma_{[1]}^2} < \frac{f\hat{\sigma}_{(i)}^2}{\sigma_{[i]}^2} \frac{\sigma_{[i]}^2}{\sigma_{[1]}^2}, i = 2, 3, \ldots, k\right)$$

$$\geqslant P(\chi_1^2(f) < \chi_i^2(f) \ r^*, i = 2, 3, \ldots, k), \tag{6.2.12}$$

where $f = n - 1$ or n depending on whether all means are unknown or all means are known, and $\chi_i^2(f)$ are independent chi-square random variables. Thus, the requirement (6.2.11) is met when

$$\int_0^\infty \left(1 - G_f\left(\frac{x}{r^*}\right)\right)^{k-1} g_f(x) \, dx \geqslant P^*, \tag{6.2.13}$$

where G_f and g_f are the distribution and density functions of the chi-square distribution with f degrees of freedom. Now f will be determined to satisfy (6.2.13), and such an f can be obtained from the tables of Gupta and Sobel (1962). From the f-value, the sample size n is determined as $f + 1$ or f depending on whether all means are unknown or all means are known.

An approximation to the f-value satisfying (6.2.13) can be obtained using the suggestion of Gupta and Sobel (1962). They suggest that the distribution of $\ln \chi^2(f)$ can be approximated by $N(\ln f, 2/f)$ distribution. Using this approximation for evaluating the probability bound of (6.2.12), the constraint (6.2.13) can be modified as

$$\int_{-\infty}^{\infty} \left\{ 1 - \Phi\left(x - \sqrt{\frac{f}{2}} \ln r^* \right) \right\}^{k-1} \phi(x)\, dx \geq P^*, \quad (6.2.14)$$

which is the same as

$$\int_{-\infty}^{\infty} \Phi^{k-1}\left(x + \sqrt{\frac{f}{2}} \ln r^* \right) \phi(x)\, dx \geq P^*. \quad (6.2.15)$$

Thus f must be chosen so that

$$f \geq 2\left(\frac{\lambda}{\ln r^*} \right)^2 = f^*, \quad (6.2.16)$$

where λ is the solution of (6.2.7). Hence the following:

Theorem 6.2.2   *The common sample size required for selecting the normal distribution with the smallest variance so that the $P(cs) \geq P^*$ whenever $\sigma_{[1]}^2 \leq r^* \ \sigma_{[2]}^2$ is $f + 1$ or f depending on whether all means are unknown or known, where f is the smallest positive integer satisfying (6.2.13). An approximation to the common sample size n is $[f^*] + 2$ or $[f^*] + 1$, where f^* is given by (6.2.16).*

Example 6.2.2.1   Let $k = 3$, $P^* = 0.99$, $r^* = 1.5$. Assume that the three means are unknown. Here $\lambda = 3.62$ so that

$$f^* = 2\left(\frac{3.62}{\ln 1.5} \right)^2 = 159.4.$$

Thus, an approximation to n, the required sample size is 161.

6.3   Selecting the Best Bernoulli Distribution

The observations from π_i follow the Bernoulli distribution with the probability function

$$f(x; \theta_i) = \theta_i^x (1 - \theta_i)^{1-x}, \ x = 0, 1, \quad (6.3.1)$$

for $i = 1, 2, \ldots, k$. The problem is the selection of the best population, which has the largest value for θ.

Let $\hat{\theta}_i$ be the mean of the sample of size n from π_i, $i = 1, 2, \ldots, k$. The ordered values of $\hat{\theta}_i$ are

$$\hat{\theta}_{[1]} \leqslant \hat{\theta}_{[2]} \leqslant \cdots \leqslant \hat{\theta}_{[k]}. \tag{6.3.2}$$

The selection procedure is to select the population associated with $\hat{\theta}_{[k]}$. If there are ties for the largest $\hat{\theta}$-value, they must be broken at random. As in the case of normal distributions, one likes to choose the common sample size, n, such that $P(cs) \geqslant P^*$ whenever $\theta_{[k]} - \theta_{[k-1]} \geqslant \delta^*$, where

$$\theta_{[1]} \leqslant \theta_{[2]} \leqslant \cdots \leqslant \theta_{[k]} \tag{6.3.3}$$

are the ordered θ-values, and δ^* and P^* are positive constants. Further, $P^* > 1/k$. The requirement on n is satisfied if

$$\min P(cs) \geqslant P^*, \tag{6.3.4}$$

where the minimum is over all parameter values such that $\theta_{[k]} - \theta_{[k-1]} \geqslant \delta^*$. The minimum of $P(cs)$ can be approximated by

$$P(\arcsin \sqrt{\hat{\theta}_{(k)}} > \arcsin \sqrt{\hat{\theta}_{(i)}},$$

$$i = 1, 2, \ldots, k - 1 \,|\, \theta_{[1]} = \cdots = \theta_{[k-1]} = \theta_{[k]} - \delta^*), \tag{6.3.5}$$

where $\hat{\theta}_{(i)}$ is the mean of the sample from the population with the parameter $\theta_{[i]}$, $i = 1, 2, \ldots, k$. The minimum of this probability occurs when $\theta_{[1]} = \cdots = \theta_{[k-1]} = (1 - \delta^*)/2$ and $\theta_{[k]} = (1 + \delta^*)/2$. Now the common sample size is chosen so that this minimum probability is at least P^*. The minimum probability is computed by using the approximate normal distribution of $\arcsin \sqrt{\hat{\theta}_{(i)}}$. Thus, n is to be chosen so that

$$\int_{-\infty}^{\infty} \Phi^{k-1}(x + \Delta^* \sqrt{4n}) \phi(x) \, dx \geqslant P^*, \tag{6.3.6}$$

where

$$\Delta^* = \arcsin \sqrt{\frac{1}{2} + \frac{\delta^*}{2}} - \arcsin \sqrt{\frac{1}{2} - \frac{\delta^*}{2}}. \tag{6.3.7}$$

This leads to

$$n \geq \left(\frac{\lambda}{2\Delta^*}\right)^2 \tag{6.3.8}$$

where λ is the solution of (6.2.7). Thus, the following is established.

Theorem 6.3.1 *An approximation to the common sample size n, required to select the best Bernoulli distribution so that $P(cs) \geq P^*$ whenever $\theta_{[k]} - \theta_{[k-1]} \geq \delta^*$ is $[n^*] + 1$, where n^* is given by (6.3.8).*

Example 6.3.1.1 Let $k = 3$, $P^* = 0.99$, and $\delta^* = 0.10$. Then $\lambda = 3.62$, $\Delta^* = \arcsin \sqrt{0.55} - \arcsin \sqrt{0.45} = 0.14$, and

$$n^* = \left(\frac{3.62}{2(0.14)}\right)^2 = 167.15$$

Thus, the approximate required common sample size is 168.

6.4   Selecting the Best Exponential Distribution

The distribution of the observations from π_i is the $NE(0, \theta_i)$ with the density function

$$f(x; \theta_i) = \frac{1}{\theta_i} e^{-x/\theta_i}, \qquad x > 0 \tag{6.4.1}$$

for $i = 1, 2, \ldots, k$. The population associated with the largest θ_i is the best population. Let the ordered θ-values be

$$\theta_{[1]} \leq \theta_{[2]} \leq \cdots \leq \theta_{[k]}. \tag{6.4.2}$$

Let $\bar{X}_i$ be the mean of the sample of size n from π_i, $i = 1, 2, \ldots, k$. The ordered sample means are

$$\bar{X}_{[1]} \leq \bar{X}_{[2]} \leq \cdots \leq \bar{X}_{[k]}. \tag{6.4.3}$$

As before, the selection rule is to choose the population corresponding to $\bar{X}_{[k]}$. If there are ties to the largest value of $\bar{X}$, ties are broken at random. The problem is to determine the common sample size n such that $P(cs) \geq P^*$ whenever $(\theta_{[k]}/\theta_{[k-1]}) \geq r^*(>1)$. If $\bar{X}_{(i)}$ is the mean of the sample from the population with the parameter $\theta_{[i]}$, $i = 1, 2, \ldots, k$, then

$$P(cs) = P(\bar{X}_{(k)} > \bar{X}_{(i)}, i = 1, 2, \ldots, k - 1). \qquad (6.4.4)$$

The n is chosen so that the minimum of $P(cs)$ is at least P^*. In other words, n is chosen so that

$$\int_0^\infty G_{2n}^{k-1}(xr^*)g_{2n}(x)\,dx \geq P^*, \qquad (6.4.5)$$

where G_f and g_f are the distribution and the density functions of the χ^2-distribution with f degrees of freedom. This n-value can be obtained from the table of Gupta (1963).

An approximation to the n-value that satisfies (6.4.5) can be derived as in Section 6.2.3 and is

$$n \geq \left(\frac{\lambda}{\ln r^*}\right)^2 = n^*, \qquad (6.4.6)$$

where λ is the solution of (6.2.7). Thus the following result:

Theorem 6.4.1   *The common sample size required to select the best exponential distribution so that $P(cs) \geq P^*$, whenever $(\theta_{[k]}/\theta_{[k-1]}) \geq r^*$ is the smallest positive integer n satisfying (6.4.5). An approximation to this value is $[n^*] + 1$, where n^* is given by (6.4.6).*

Example 6.4.1.1   Let $k = 3$, $P^* = 0.99$, $r^* = 1.5$. Here $\lambda = 3.62$ and $n^* = (3.62/\ln 1.5)^2 = 79.7$. Thus an approximation to the required sample size is 80.

Chapter 7

Biomedical Experiments and Clinical Trials

7.1 Introduction

In clinical trials as well as in some biomedical experiments, the sample size determination is usually posed in relation to tests of hypotheses. However, in these experiments data collection is associated with special problems. Patients may drop out of the study; animals may die due to unknown causes. These factors lead to a certain incompleteness in the observations from some experimental units. The methodology developed in earlier chapters tacitly assumed that the observations are complete. As such in the present context, new techniques are needed. Further, some special experimental setups that were not considered earlier will be discussed here. The two sections deal with discrete and continuous data situations.

7.2 Experiments with Qualitative Response Variable

Here experiments that are undertaken with a view to compare the effects of two treatments will be considered. Further, the response variable is a qualitative one. The possible values are finite and when there are only two values, they are coded as 0 and 1. The basic null hypothesis is that the two treatments are equally effective. Tests with prescribed α level will be given, and sample sizes will be determined so that these tests have prescribed power at given alternatives.

7.2.1 McNemar's Test

Consider the following two types of experimental settings:

(i) n matched pairs of experimental units (humans, animals, etc.) are included in the experiment. One unit in each pair is randomly assigned to treatment A and the other to treatment B. For each unit $0 - 1$ response is noted.

(ii) Each of the n experimental units are given both treatments A and B. Randomly chosen n_1 units are given the treatment sequence A followed by B and the remaining $n - n_1$ units are given the treatment sequence B followed by A. Such experiments are known as cross-over or change-over designs. A $0 - 1$ response is noted for each subject under each treatment. An adequate rest period between the treatments should be used so that carry-over effects are washed out.

In both of these settings the results can be summarized in terms of n_{ij}, where n_{ij} is the number of pairs (or units) with response i for A and j for B, $i, j = 0, 1$. The summary table is the following:

		Treatment B	
		0	1
Treatment A	0	n_{00}	n_{01}
	1	n_{10}	n_{11}

The cell frequencies n_{ij} follow a multinomial distribution with parameters θ_{ij}, $i, j = 0, 1$; and n. The null hypothesis of interest is that the two treatments are equally effective, that is,

$$H_0: \theta_{10} + \theta_{11} = \theta_{01} + \theta_{11}, \qquad (7.2.1)$$

or equivalently,

$$H_0: \theta_{10} = \theta_{01}, \qquad (7.2.2)$$

and the alternative hypothesis is $H_A: \theta_{10} > \theta_{01}$. In such experiments, one treatment is usually the standard or control, whereas the other is the experimental treatment which is expected to perform better than the standard. It is natural to consider the test with the critical region

$$\frac{n_{10}}{n} - \frac{n_{01}}{n} > c, \tag{7.2.3}$$

where the constant c is chosen so that the test is an α-level test. Thus, c must satisfy

$$P(n_{10} - n_{01} > cn \mid H_0) \leqslant \alpha. \tag{7.2.4}$$

Let $S = n_{10} + n_{01}$, so that $n_{10} - n_{01} = 2(n_{10} - S/2)$. Further, the conditional distribution of n_{10} given S is the binomial distribution with parameters S and $\theta_{10}/(\theta_{10} + \theta_{01}) = \theta_{10}/\psi$, say. The condition (7.2.4) will be met if

$$P\left(n_{10} - \frac{S}{2} > \frac{cn}{2} \,\middle|\, S, H_0\right) \leqslant \alpha \tag{7.2.5}$$

for each value of S. This in turn can be approximately satisfied by choosing $cn/2 = z_\alpha \sqrt{(S/4)}$ when n is large. Thus, an approximate α-level test is given by the critical region

$$n_{10} > \frac{S}{2} + z_\alpha \sqrt{\frac{S}{4}}. \tag{7.2.6}$$

Now if one wants to control the power at $1 - \beta$ for $\theta_{10} - \theta_{01} = \delta^* > 0$, n must be chosen so that

$$P\left(n_{10} > \frac{S}{2} + z_\alpha \sqrt{\frac{S}{4}} \,\middle|\, \theta_{10} - \theta_{01} = \delta^*\right) = 1 - \beta. \tag{7.2.7}$$

The probability on the left side of (7.2.7) can be approximated as follows:

$$E_S P\left(n_{10} > \frac{S}{2} + z_\alpha \sqrt{\frac{S}{4}} \,\middle|\, S, \qquad \theta_{10} - \theta_{01} = \delta^*\right) \tag{7.2.8}$$

$$\doteq E_S\left[1 - \Phi\left\{\left(\frac{S}{2} + z_\alpha \sqrt{\frac{S}{4}} - \frac{S\theta_{10}}{\psi}\right) \,\middle/\, \sqrt{S\frac{\theta_{10}}{\psi}\left(1 - \frac{\theta_{10}}{\psi}\right)}\right\}\right]$$

by approximating the conditional binomial distribution of n_{10}, given S, with a normal distribution. This expression in turn can be approximated by

$$1 - \Phi\left[\left\{E_1(S)\left(\frac{1}{2} - \frac{\theta_{10}}{\psi}\right) + \frac{z_\alpha}{2}\sqrt{E_1(S)}\right\} \Big/ \sqrt{\frac{\theta_{10}}{\psi}\left(1 - \frac{\theta_{10}}{\psi}\right)E_1(S)}\right],$$

$$(7.2.9)$$

where $E_1(S) = E(S|\theta_{10} - \theta_{01} = \delta^*) = n(\theta_{10} + \theta_{01}) = n\psi$. By equating the approximate power given in (7.2.9) to $1 - \beta$, the equation that gives an approximation to n satisfying (7.2.7) is

$$n\psi\left(\frac{1}{2} - \frac{\theta_{10}}{\psi}\right) + \frac{z_\alpha}{2}\sqrt{n\psi} = -z_\beta\sqrt{\theta_{10}\left(1 - \frac{\theta_{10}}{\psi}\right)n}, \quad (7.2.10)$$

which is the same as

$$\sqrt{n}(2\theta_{10} - \psi) = z_\alpha\sqrt{\psi} + z_\beta\sqrt{4\theta_{10}\left(1 - \frac{\theta_{10}}{\psi}\right)} \quad (7.2.11)$$

simplifying to

$$\sqrt{n}\,\delta^* = z_\alpha\sqrt{\psi} + z_\beta\sqrt{\frac{4\theta_{10}\theta_{01}}{\psi}}$$

$$= z_\alpha\sqrt{\psi} + z_\beta\sqrt{\frac{(\psi + \delta^*)(\psi - \delta^*)}{\psi}}.$$

$$(7.2.12)$$

It may be noted that one also needs to specify ψ in order to solve (7.2.12) for n. Thus the following results:

Theorem 7.2.1 *An approximate sample size needed for an approximate α-level one sided test of H_0: $\theta_{10} = \theta_{01}$ with a power of $1 - \beta$ at $\theta_{10} - \theta_{01} = \delta^*(>0)$ and $\theta_{10} + \theta_{01} = \psi$ is $[n^*] + 1$, where n^* satisfies (7.2.12).*

Consider the following example:

Example 7.2.1.1 A cardiologist is interested in testing new heart medication (A) against a standard drug (B) in controlling angina. He chooses n matched pairs of individuals and in each pair randomly assigns treatments A and B to individual patients. Cross-over designs with wash-out period is not practical here as the patient cannot be left untreated for any specific period. The responses noted were no angina (1) and angina on at least one occasion (0) over a one-week period. He expects $\psi = 0.5$ and wants to use a 0.05 level one-sided test to establish that the medication A controls angina better than medication B. If he wishes to have a power of 0.9, when $\delta^* = 0.1$, the required n from (7.2.12) is

$$n = \left| \left\{ \frac{1.645\sqrt{0.5} + 1.282\sqrt{(0.6)(0.4)/0.5}}{0.1} \right\}^2 \right| + 1 = 421.$$

Thus he needs 421 matched pairs of patients for the study.

It may be noted that the approximation considered in Theorem 7.2.1 was given by Lachin (1981). This was also derived by Lee *et al.* (1984) using a slightly different approach. Sample size tables based on the unconditional power function are given by Schork and Williams (1980). This problem was also considered by Connor (1987), who gave an alternative approximation to the sample size.

When the alternative hypothesis is two sided, that is, H_A: $\theta_{10} \neq \theta_{01}$, the critical region for the usual McNemar's test is

$$T = \frac{(n_{10} - n_{01})^2}{(n_{10} + n_{01})} > \chi^2_{1-\alpha}(1). \qquad (7.2.13)$$

In order to control the power at $1 - \beta$ for the alternative $|\theta_{10} - \theta_{01}| = \delta^*$, one needs to choose n so that

$$P(T > \chi^2_{1-\alpha}(1) \mid \delta^*) = 1 - \beta. \qquad (7.2.14)$$

Under this alternative $T \sim \chi^2(1, \Delta)$, where

$$\Delta = E(T) - 1$$

$$= \frac{n\delta^{*2}}{\psi} - \frac{\delta^{*2}}{\psi^2}. \qquad (7.2.15)$$

Thus the condition (7.2.14) can be viewed as

$$P(\chi^2(1; \Delta) > \chi^2_{1-\alpha}(1)) = 1 - \beta. \qquad (7.2.16)$$

From the table of noncentral χ^2-distribution, one can determine the required n. Approximating the distribution of

$$Z = \frac{\sqrt{2\chi^2(\nu;\Delta)} - \sqrt{2(\nu + \Delta^2) - \dfrac{\nu + 2\Delta^2}{\nu + \Delta^2}}}{\sqrt{\dfrac{\nu + 2\Delta^2}{\nu + \Delta^2}}} \qquad (7.2.17)$$

by $N(0, 1)$ distribution, an approximation to n satisfying (7.2.16) is $[n^*] + 1$, where

$$n^* = \frac{\psi}{\delta^{*2}} (z_{\alpha/2} + z_\beta)^2 + \frac{1}{\psi}. \qquad (7.2.18)$$

Thus,

Theorem 7.2.2 *An approximate sample size needed for an α-level test of H_0: $\theta_{10} = \theta_{01}$ against the alternative H_A: $\theta_{10} \neq \theta_{01}$ with a power of $1 - \beta$, at $|\theta_{10} - \theta_{01}| = \delta^*$ and $\psi = \theta_{10} + \theta_{01}$ is $[n^*] + 1$, where n^* is given by (7.2.18).*

Example 7.2.2.1 For a two-sided test with specifications as given in Example 7.2.1.1, one gets

$$n^* = \frac{0.5}{(0.1)^2} (1.96 + 1.282)^2 + \frac{1}{0.5}$$

$$= 527.5$$

Thus, the required number of matched pairs for this test are 528.

The problem of bioequivalence with $0 - 1$ response variable was considered by Dunnett and Gent (1977), and the necessary sample size discussion was given by Rodary *et al.* (1989).

7.2.2   Testing Equality of r Treatment Effects with c Categorical Responses

Consider an experiment with r treatments, where n_i subjects receive the ith treatment, $i = 1, 2, \ldots, r$. The response variable can take c values. Let $n = \Sigma_{i=1}^{r} n_i$ and $Q_i = n_i/n$, $i = 1, 2, \ldots, r$. The data can be summarized in a $r \times c$ table:

		\multicolumn{6}{c}{Response Categories}				
		1	2	...	j	...	c	Total
	1	n_{11}	n_{12}	...	n_{1j}	...	n_{1c}	n_1
	2	n_{21}	n_{22}	...	n_{2j}	...	n_{2c}	n_2
	.	.	.	...	.	...	.	.
Treatments	.	.	.	...	.	...	.	.
	.	.	.	...	.	...	.	.
	i	n_{i1}	n_{i2}	...	n_{ij}	...	n_{ic}	n_i
	.	.	.	...	.	...	.	.
	.							
	.							
	r	n_{r1}	n_{r2}	...	n_{rj}	...	n_{rc}	n_r

where n_{ij} is the number of subjects with j^{th} response category who received i^{th} treatment, $i = 1, 2, \ldots, r$; $j = 1, 2, \ldots, c$. Let θ_{ij} be the probability of getting j^{th} response under treatment i, $i = 1, 2, \ldots, r$; $j = 1, 2, \ldots, c$, so that $\Sigma_{j=1}^{c} \theta_{ij} = 1$ for every i. The null hypothesis of interest is that all treatments have the same effect, that is,

$$H_0: \theta_{ij} = \alpha_j, \qquad j = 1, 2, \ldots, c \text{ for each } i. \qquad (7.2.19)$$

Letting $\hat{\alpha}_j = \Sigma_{i=1}^{r} n_{ij}/n$, the critical region of the usual χ^2-test of H_0 is

$$T = \sum_{i,j} \frac{(n_{ij} - n_i \hat{\alpha}_j)^2}{n_i \hat{\alpha}_j} > \chi^2_{1-\alpha}(\nu), \qquad (7.2.20)$$

where $\nu = (r - 1)(c - 1)$. If one wants to have a power of $1 - \beta$ at the alternative set of parameters $\theta_{ij} = \theta'_{ij} \equiv \alpha_j + \delta_{ij}$, for specified δ_{ij}, the sample sizes are to be chosen so that

$$P(T > \chi^2_{1-\alpha}(\nu) \mid \theta'_{ij}) = 1 - \beta. \qquad (7.2.21)$$

Under the alternative, T will be approximately a noncentral χ^2-variable with ν degrees of freedom and the noncentrality parameter Δ [cf. Lachin (1977)], where

$$\Delta^2 = n\Sigma_j \frac{1}{\alpha_j} \{\Sigma_i Q_i \delta_{ij}^2 - (\Sigma_i Q_i \delta_{ij})^2\}, \text{ for } r \geq 2, c \geq 2,$$

$$= nQ_1 Q_2 \left\{ \Sigma_j \frac{(\delta_{1j} - \delta_{2j})^2}{\alpha_j} \right\}, \text{ for } r = 2, c \geq 2, \qquad (7.2.22)$$

$$= n\left\{ \Sigma_i Q_i \delta_{i1}^2 - \frac{(\Sigma_i Q_i \delta_{i1})^2}{\alpha_1(1 - \alpha_1)} \right\}, \text{ for } r \geq 2, c = 2.$$

The noncentrality parameter (7.2.22) can be obtained by putting $E(n_{ij}) = n_i \theta'_{ij}$, $E(\hat{\alpha}_j) = \alpha_j + \Sigma_i n_i \delta_{ij}/n$ in the numerator and $\hat{\alpha}_j \doteq \alpha_j$ in the denominator of T of (7.2.20). Thus, from (7.2.21), one gets

$$P(\chi^2(\nu, \Delta) > \chi^2_{1-\alpha}(\nu)) = 1 - \beta. \qquad (7.2.23)$$

Using the noncentral χ^2 distribution tables, for given $\{\alpha_j\}$ and $\{\delta_{ij}\}$, the required sample sizes n_i can be determined.
    The previous ideas will now be illustrated.

Example 7.2.2.1   Consider an experiment planned to test the equally effective hypothesis for four treatments with dichotomous responses using a 0.05 level test. The experimenter wants to have a power of 0.90 to detect the differences when

$$\theta'_{11} = 0.55, \qquad \theta'_{12} = 0.45$$
$$\theta'_{21} = 0.45, \qquad \theta'_{22} = 0.55$$
$$\theta'_{31} = 0.50, \qquad \theta'_{32} = 0.50$$
$$\theta'_{41} = 0.50, \qquad \theta'_{42} = 0.50,$$

and he wants to use $n_1 = n_2 = n_3 = n_4 = m$, so that $n = 4m$. From the noncentral χ^2 table of Haynam et al. (1973), the noncentrality for a χ^2 with 3 d.f., $\alpha = .05$ and power $= 0.9$, is 12.654 and from (7.2.22) it is .02m. Thus,

$$.02\,m = 12.654$$
$$m = 632.7.$$

A total of 2532 subjects are needed for this experiment in which each treatment is administered on 633 patients.

An approximation to the required sample size can be obtained by using the normal approximation as outlined in Section 7.2.1.

7.2.3 Long-Term Trials

Here the sample sizes are determined when a control treatment and an experimental treatment are compared over a long period of time for an event rate. Due to the long duration of study, some subjects may leave the study due to various reasons. The problem is to determine the sample size for testing the equality of event rates controlling the power at a given alternative. This problem was discussed in Chapter 2 without considering the possibility of dropout of subjects from the study. If R is the expected drop-out rate, and if n is the common sample size calculated assuming no drop-outs, the recommended sample size required with drop-outs, n_d, [see Lachin (1981)] is

$$n_d = \frac{n}{(1 - R)^2}. \tag{7.2.24}$$

For further discussion on long-term trials, the interested reader is referred to Halperin et al. (1968), Wu et al. (1980), and Lakatos (1986).

7.2.4 Case Control Studies

In case-control studies, individuals with a particular condition or disease (the cases) and individuals without the condition or disease (the controls) are compared for studying the effect of risk factors relevant to the development of the condition or disease under study. In this section, studies of one risk factor where stratification using levels of a confounding variable are considered.

The data in the jth stratum $(j = 1, 2, \ldots, J)$ based on the level of the confounding variable can be presented in the following table:

Risk group	Case	Control	Total
Exposed	a_j	b_j	m_{1j}
Not exposed	c_j	d_j	m_{2j}
Total	n_{1j}	n_{2j}	n_j

Let $n = \Sigma_{j=1}^{J} n_j$, $t_j = n_j/n$, and $s_j = n_{1j}/n_j$. Thus, t_j denotes the relative size of the jth stratum and s_j denotes the proportion of cases in the jth stratum. Let π_{1j} and π_{2j} be the probabilities with which the individuals are exposed to the risk in the case and control groups, respectively, in the jth stratum, $j = 1, 2, \ldots, J$. Then the well-known odds ratio between exposure and disease in stratum j is

$$\theta_j = \frac{\dfrac{\pi_{1j}}{(1 - \pi_{1j})}}{\dfrac{\pi_{2j}}{(1 - \pi_{2j})}}, \qquad j = 1, 2, \ldots, J. \qquad (7.2.25)$$

It is assumed that $\theta_1 = \theta_2 = \cdots = \theta_J (= \theta)$, and the problem is to test the null hypothesis H_0: $\theta = 1$ against the alternative hypothesis H_A: $\theta > 1$. This null hypothesis is equivalent to $\pi_{1j} = \pi_{2j}$, $j = 1, 2, \ldots, J$. One estimates π_{1j} and π_{2j} by $p_{1j} = a_j/n_{1j}$, and $p_{2j} = b_j/n_{2j}$, respectively. The test statistic suggested by Cochran (1954) for this problem is

$$C_0 = \frac{\Sigma_{j=1}^{J} w_j (p_{1j} - p_{2j})}{\sqrt{\Sigma_{j=1}^{J} w_j \bar{p}_j \bar{q}_j}}, \qquad (7.2.26)$$

where

$$w_j = \frac{n_{1j} n_{2j}}{n_{1j} + n_{2j}},$$
$$\bar{p}_j = \frac{n_{1j} p_{1j} + n_{2j} p_{2j}}{n_j}, \qquad \bar{q}_j = 1 - \bar{p}_j. \qquad (7.2.27)$$

Under H_0, C_0 is asymptotically distributed as a $N(0, 1)$ variable, and hence the critical region for an α-level test is

$$C_0 > z_\alpha. \tag{7.2.28}$$

Under H_A, $\pi_{1j} > \pi_{2j}$, for at least one j, the variable

$$C = \frac{\sum_{j=1}^J w_j\{(p_{1j} - p_{2j}) - (\pi_{1j} - \pi_{2j})\}}{\sqrt{\sum_{j=1}^J w_j^2\left\{\dfrac{\pi_{1j}(1 - \pi_{1j})}{n_{1j}} + \dfrac{\pi_{2j}(1 - \pi_{2j})}{n_{2j}}\right\}}} \tag{7.2.29}$$

is asymptotically distributed as $N(0, 1)$.

Now if one wants to control the power at $1 - \beta$ for $\theta = \theta^*$ (>1), the sample size n must be chosen such that

$$P(C_0 > z_\alpha \mid \theta^*) = 1 - \beta. \tag{7.2.30}$$

For every j, π_{1j}, π_{2j} are related to θ^* through the equation

$$\pi_{1j} = \frac{\pi_{2j}\theta^*}{1 - \pi_{2j} + \pi_{2j}\theta^*}. \tag{7.2.31}$$

Usually there will be prior knowledge on $\{\pi_{2j}\}$, which combined with the specified θ^* value, determines $\{\pi_{1j}\}$ values. For such configuration of $\{\pi_{1j}\}$ values, Equation (7.2.30) in conjunction with asymptotic normal distribution of the variable C_0 yields

$$1 - \Phi\left(\frac{z_\alpha\sqrt{\sum_{j=1}^J w_j \bar{p}_j \bar{q}_j} - \sum_{j=1}^J w_j(\pi_{1j} - \pi_{2j})}{\sqrt{\sum_{j=1}^J w_j^2\left\{\dfrac{\pi_{1j}(1 - \pi_{1j})}{n_{1j}} + \dfrac{\pi_{2j}(1 - \pi_{2j})}{n_{2j}}\right\}}}\right) = 1 - \beta. \tag{7.2.32}$$

Since $\bar{p}_j$ and $\bar{q}_j$ are unobserved random variables, they will be replaced by their expectations $\bar{\pi}_j = s_j\pi_{1j} + (1 - s_j)\pi_{2j}$ and $1 - \bar{\pi}_j$, respectively. Thus, the requirement (7.2.32) implies

$$\frac{z_\alpha\sqrt{\sum_{j=1}^J w_j \bar{\pi}_j(1 - \bar{\pi}_j)} - \sum_{j=1}^J w_j(\pi_{1j} - \pi_{2j})}{\sqrt{\sum_{j=1}^J w_j^2\left\{\dfrac{\pi_{1j}(1 - \pi_{1j})}{n_{1j}} + \dfrac{\pi_{2j}(1 - \pi_{2j})}{n_{2j}}\right\}}} = -z_\beta. \tag{7.2.33}$$

For pre-chosen design parameters $\{s_j\}$ and $\{t_j\}$, the equation can be rewritten in terms of n and can be solved for n. The solution is

$$n^* = \frac{[z_\alpha\sqrt{\Sigma_{j=1}^{J} t_j s_j(1-s_j)\bar{\pi}_j(1-\bar{\pi}_j)} + z_\beta\sqrt{\Sigma_{j=1}^{J} t_j s_j(1-s_j)\{(1-s_j)\pi_{1j}(1-\pi_{1j}) + s_j\pi_{2j}(1-\pi_{2j})\}}]^2}{[\Sigma_{j=1}^{J} t_j s_j(1-s_j)(\pi_{1j}-\pi_{2j})]^2}.$$

$$(7.2.34)$$

The required sample size is $[n^*] + 1$. For an example and further discussion on this problem, the reader is referred to Woolson *et al.* (1986).

Sample size determination with more than one risk factor in a case-control study was given by Rao (1986).

7.2.5   Two-Stage Designs

In clinical trials the data are usually reviewed at least once before the completion of the study. In this review, one can reach a decision whether to continue or terminate the study. This process leads to designing two-stage testing procedures of the following types:

(1) At stage one a definite conclusion is reached whether to reject or retain the formulated null hypothesis, and if no decision can be reached because of inadequate evidence, continue to stage two to reach a decision.

(2) At stage one retain H_0 and terminate the study, or reject H_0 and continue to stage two to enhance the evidence for the decision.

Such two-stage procedures using dichotomous response variables were recently studied by Case *et al.* (1987); and Thall *et al.* (1988). Two-stage procedures for testing the effectiveness of several treatments and selecting the best treatment was considered by Thall *et al.* (1988).

7.3   Experiments with Quantitative Response Variable

In this section, comparative experiments where the response is a quantitative variable will be considered. This discussion includes the bioequivalence problem and design of clinical trials for comparison of survival distributions.

7.3.1   Testing for Bioequivalence

Let μ_E and μ_S be the mean values of a measure of biological activity for the experimental and standard formulations of a drug. The two formulations are considered to be bioequivalent if $A < \mu_E - \mu_S < B$ for specified constants A and B. An experiment with n subjects on each formulation is to be designed, so that the data will be used to test the hypothesis of bioequivalence. This problem is formulated as testing the null hypothesis

$$H_0: \mu_E - \mu_S \leq A \text{ or } \mu_E - \mu_S \geq B \qquad (7.3.1)$$

against the alternative hypothesis

$$H_A: A < \mu_E - \mu_S < B. \qquad (7.3.2)$$

The mathematical problem to be resolved is the determination of n, so that one can construct an α-level test having an appropriate power of $1 - \beta$. The response variable is assumed to follow a normal distribution with variance σ^2.

Let $\bar{X}_E$ and $\bar{X}_S$ be the sample means. The test statistic proposed by Anderson and Hauck (1983) is

$$T = \frac{\bar{X}_E - \bar{X}_S - \frac{1}{2}(A + B)}{s_p \sqrt{\frac{2}{n}}}, \qquad (7.3.3)$$

where s_p^2 is the pooled estimator of σ^2 based on $\nu = 2(n - 1)$ degrees of freedom. The statistic T has a noncentral t distribution with ν degrees of freedom and noncentrality parameter

$$\Delta = \left\{ \mu_E - \mu_S - \frac{1}{2}(A + B) \right\} \Big/ \left\{ \sigma \sqrt{\frac{2}{n}} \right\}. \qquad (7.3.4)$$

The critical region of the test is

$$|T| < C, \qquad (7.3.5)$$

where c can be determined such that

$$P(|T| < c \,|\, \mu_E - \mu_S = A) = P(|T| < c \,|\, \mu_E - \mu_S = B) = \alpha. \qquad (7.3.6)$$

Since determination of c so that the test has size α is difficult, it is suggested to compute the p-value and draw appropriate conclusions about equivalence. An approximation to the p-value at $\mu_E - \mu_S = B$ is

$$\rho = F_\nu(|T_{\text{obs}}| - d) - F_\nu(-|T_{\text{obs}}| - d), \qquad (7.3.7)$$

where

$$d = \frac{B - A}{2s_p \sqrt{\dfrac{2}{n}}}, \qquad (7.3.8)$$

and $F_\nu(\cdot)$ is the distribution function of $T(\nu)$ distribution. Thus the formulations will be declared as bioequivalent if $\rho \leqslant \alpha$.

The power calculation is somewhat simpler when $\mu_E - \mu_S = (A + B)/2$. So the power at this configuration is equated to $1 - \beta$, and the common sample size is determined. In other words, n is chosen so that

$$P\left(|T| < c \,\Big|\, \mu_E - \mu_S = \frac{A + B}{2}, \sigma\right) = 1 - \beta. \qquad (7.3.9)$$

This will be satisfied if $c = t_{\beta/2}(\nu)$, the upper $100(\beta/2)$ percentile of the $T(\nu)$ distribution. However, c must also satisfy the condition,

$$P(T \leqslant c \,|\, \mu_E - \mu_S = B, \sigma) - P(T \leqslant -c \,|\, \mu_E - \mu_S = B, \sigma) = \alpha. \qquad (7.3.10)$$

Using the fact that $c = t_{\beta/2}(\nu)$, Equation (7.3.10) can be rewritten as

$$P(T(\nu, \Delta_1) \leqslant t_{\beta/2}(\nu)) - P(T(\nu, \Delta_1) \leqslant -t_{\beta/2}(\nu)) = \alpha, \qquad (7.3.11)$$

where $\Delta_1 = 1/2\,(B - A)/\{\sigma\sqrt{2/n}\}$, $\nu = 2\,(n - 1)$. Solving this equation iteratively for $\Delta_1 = \tilde{\Delta}_1$, and then n is obtained as

$$\tilde{n} = \frac{8\,\tilde{\Delta}_1^2 \tilde{\sigma}^2}{(B - A)^2}, \qquad (7.3.12)$$

where $\tilde{\sigma}^2$ is a prior estimate of σ^2.

Approximating the noncentral t distribution by central distribution and ignoring the second term on the left-hand side, Equation (7.3.11) simplifies to

$$P\left(T(\nu) \leqslant t_{\beta/2}(\nu) - \frac{(B-A)}{2\sigma\sqrt{\frac{2}{n}}}\right) = \alpha. \qquad (7.3.13)$$

This equation is satisfied when

$$n = \frac{8(t_{\beta/2}(\nu) + t_\alpha(\nu))^2 \sigma^2}{(B-A)^2}, \qquad (7.3.14)$$

where $\nu = 2(n-1)$. One can solve this equation iteratively to get the final solution n^*. An approximation to the required sample size is $[n^*] + 1$. Thus,

Theorem 7.3.1 *An approximation to the common sample size for an α-level test of H_0: $\mu_E - \mu_S \leqslant A$ or $\mu_E - \mu_S \geqslant B$ against the alternative H_A: $A < \mu_E - \mu_S < B$, giving a power of $1 - \beta$ at $\mu_E - \mu_S = (A+B)/2$ is $[n^*] + 1$, where n^* is the solution of (7.3.14).*

Note that $(t_{\beta/2}(\nu) + t_\alpha(\nu))$ in (7.3.14) is an approximation to $\tilde{\Delta}_1$ given in (7.3.12).

Example 7.3.1.1 Consider the example discussed by Anderson and Hauck (1983), in which $A = \log 0.75 = -0.12494$, $B = \log 1.25 = 0.09691$, $\sigma^2 = 0.012$, $\alpha = 0.05$ and $1 - \beta = 0.80$. Initially choosing $\nu = \infty$,

$$n = \frac{8(1.282 + 1.645)^2 (0.012)}{(0.09691 + 0.12494)^2} = 16.71.$$

Now take $\nu = 2(16.71 - 1) = 31.42 \doteq 32$ and compute the right-hand side expression of (7.3.14) to get $n = 17.57$. A reasonable approximation to n is $[17.57] + 1 = 18$, which was also given in the cited reference.

7.3.2 Two Period, Two Treatment, Crossover Study

A commonly used experimental setting in clinical trials is the crossover design. A brief reference to this study with $0 - 1$

response variable was made in Section 7.2.1. A detailed discussion of this design with a continuous response variable will be undertaken here.

Let A and B be two treatments administered on $2n$ randomly selected subjects. The $2n$ subjects will be randomly divided into two groups of n each. The treatment sequence A followed by B will be tested on each subject of the first group, and the sequence B followed by A will be tested on each subject of the second group. The experimental setup will then appear as the following:

Periods	1	2	Subjects ...	n	$n+1$	$n+2$	...	$2n$
I	A	A	...	A	B	B	...	B
II	B	B	...	B	A	A	...	A

Let Y_{ijk} be the response in the kth period on the jth subject in the ith sequence; $i = 1, 2; j = 1, 2, \ldots, n; k = 1, 2$. Let $\mu, \rho_i, \tau_i, \delta_i, \gamma_{ij}$ and e_{ijk} denote the general mean, i^{th} period effect, i^{th} treatment direct effect, i^{th} treatment residual effect, j^{th} subject effect in the ith sequence and random errors, respectively. Now the usual model is

$$Y_{1j1} = \mu + \rho_1 + \gamma_{1j} + \tau_A + e_{1j1}$$
$$Y_{2j1} = \mu + \rho_1 + \gamma_{2j} + \tau_B + e_{2j1}$$
$$Y_{1j2} = \mu + \rho_2 + \gamma_{1j} + \tau_B + \delta_A + e_{1j2}$$
$$Y_{2j2} = \mu + \rho_2 + \gamma_{2j} + \tau_A + \delta_B + e_{2j2}, \qquad j = 1, 2, \ldots, n.$$

(7.3.15)

It is of interest to determine the common sample size by controlling the power of an α-level test of H_0: $\tau_A = \tau_B$. To this end, it is assumed that $\delta_A = \delta_B$, that is, the residual effects of both treatments are same. Let $Z_{ij} = Y_{ij1} - Y_{ij2}$; $i = 1, 2; j = 1, 2, \ldots, n$. A test of H_0: $\tau_A = \tau_B$ is equivalent to the two-sample problem in relation to the two independent samples $\{Z_{1j}\}$ and $\{Z_{2j}\}$.

Making the distributional assumption that e_{ijk} are independently and normally distributed with mean zero and variance σ^2, γ_{ij} are independently and normally distributed with mean zero and variance σ_b^2, and e_{ijk} and γ_{ij} are independently distributed,

H_0: $\tau_A = \tau_B$ can be tested by the two-sample t-test. Following a similar argument as in Section 2.2.6, one gets

Theorem 7.3.2 *The common sample size n needed to give a power of* $1 - \beta$ *when* $|\tau_A - \tau_B| = \Delta$ *(>0) for a two-sided, α-level test of* H_0: $\tau_A = \tau_B$ *is* $(v/2 + 1)$, *where v is the smallest positive even integer satisfying*

$$n = \frac{v}{2} + 1 \doteq \frac{2(2\sigma^2)(t_{\alpha/2}(v) + t_{\beta}(v))^2}{\Delta^2}. \tag{7.3.16}$$

For a one-sided test, replace $\alpha/2$ by α in (7.3.16).

Suppose the normality assumption on e_{ijk} and γ_{ij} is not valid. Then the equality of the direct treatment effects can be tested by the Wilcoxon two-sample test, and an approximation to the common sample size can be determined following the work of Noether (1987).

7.3.3 *Clinical Trials for Comparison of Survival Distributions*

Clinical trials undertaken to compare the survival under two treatments have the following format. After the start of the study, there is an accrual period during which patients enter the trial and are randomly assigned to the treatments. The accrual period will be followed by a period called the "follow-up period" during which patients will be under observation. In these studies, times for some critical events such as death (which will be referred to as death times) for some patients are not observed and are said to be censored. In general, censoring may come about in two ways: (1) during both accrual and follow-up periods, patients may be lost to follow-up or leave the study; (2) patients may still be alive at the termination of the study. The design of such clinical trials consists of determining the length of the trial required to obtain a desired power for an α-level test, where the survival distributions, accrual rate, follow-up period and rate of loss to follow-up are specified. The problem of finding the sample size in this setting reduces to that of determining the trial length with given specifications.

The following discussion is based on the work of Rubinstein *et al.* (1981). The design problem mentioned earlier is solved by making the following assumptions:

(1) During the accrual period of T years, patients enter the trial according to a Poisson process with the rate "a" per year. It is equally likely for each patient to receive the control or experimental treatment. The follow-up period is of τ additional years.

(2) The death times are independent with common exponential distribution for each treatment. The hazard rates for the control and experimental treatment are λ_c and λ_e, respectively.

(3) The times from entry to "loss to follow-up" are independent, and they are independent of the entry times as well as of death times. These times follow exponential distribution with rate ϕ_c for the control group and ϕ_e for experimental group.

The null hypothesis to be tested is

$$H_0: \lambda_c = \lambda_e, \tag{7.3.17}$$

against the alternative hypothesis

$$H_A: \lambda_c > \lambda_e. \tag{7.3.18}$$

The accrual period length T is determined so that the usual test of H_0 in (7.3.17) has power $1 - \beta$ at the alternative $\lambda_c = r^*\lambda_e$, where $r^* > 1$.

Let D_c and D_e be the numbers of deaths in the two treatment groups. Further let T_c and T_e be the total time at risk (sum of survival times and censored times) for the control and experimental groups, respectively. The test is based on the estimators

$$\hat{\lambda}_c = \frac{D_c}{T_c}; \qquad \hat{\lambda}_e = \frac{D_e}{T_e}. \tag{7.3.19}$$

An approximate distribution of $\ln \hat{\lambda}_i$ is $N(\ln \lambda_i, 1/D_i)$, $i = c, e$. Using this asymptotic distribution, the critical region of the test is

$$\ln\left(\frac{\hat{\lambda}_c}{\hat{\lambda}_e}\right) > z_\alpha \sqrt{\frac{1}{D_c} + \frac{1}{D_e}}. \tag{7.3.20}$$

The power requirement mentioned after Equation (7.3.18) can be written as

$$
1 - \Phi \left(z_\alpha - \frac{\ln r^*}{\sqrt{\dfrac{1}{D_c} + \dfrac{1}{D_e}}} \right) = 1 - \beta,
\tag{7.3.21}
$$

and hence

$$
\frac{(\ln r^*)^2}{(z_\alpha + z_\beta)^2} = \frac{1}{D_c} + \frac{1}{D_e}.
\tag{7.3.22}
$$

Since the study is being designed, D_c and D_e are unobserved random variables. Replacing D_c and D_e by their expected values in (7.3.22), one gets

$$
\frac{(\ln r^*)^2}{(z_\alpha + z_\beta)^2} = \frac{1}{E(D_c)} + \frac{1}{E(D_e)}.
\tag{7.3.23}
$$

It can be shown that [cf. Rubinstein et al. (1981), Appendix]

$$
E(D_i) = \frac{\alpha T \lambda_i}{2\lambda_i^*} \left\{ 1 - \frac{e^{-\lambda_i^* \tau}}{\lambda_i^* T} (1 - e^{-\lambda_i^* T}) \right\},
\tag{7.3.24}
$$

where

$$
\lambda_i^* = \lambda_i + \phi_i, \qquad i = c, e.
\tag{7.3.25}
$$

Using the expression (7.3.24) for $E(D_i)$ in (7.3.23), the resulting equation will be solved for T. Tables for $T + \tau$ for $\phi_i = 0$ and $\phi_i = \lambda_i/4$ are given by Rubinstein et al. (1981).

For an alternative formulation of this problem, one is referred to George and Desu (1974).

From a simulation study, Rubinstein et al. (1981) observed that "the above method gives trial lengths which yield accurate power when used with the nonparametric Mantel-Haenszel (logrank) test." When there is no loss to follow-up and no follow-up period, trial lengths obtained by Rubinstein et al. differ very little from the results of George and Desu (1974).

In some clinical trials the objective is to compare three or more treatment groups. Makuch and Simon (1982) considered the problem of determining sample size requirements and trial lengths when one is comparing time-to-failure (i.e., death times) among k treatment groups.

Schoenfeld and Richter (1981) prepared nomograms for calculating sample sizes (the number of patients needed) for a clinical trial where two treatments are being compared, and survival is the end point. In preparing these nomograms they assumed that the survival times follow an exponential distribution, and that patients enter the study uniformly.

The logrank test is commonly used to compare two survival distributions. Schoenfeld (1981) considered the problem of sample size determination when the data is analyzed using the logrank test. Also Freedman (1982) considered this problem and tabulated the number of patients required for such studies.

Morgan (1985) noted that in clinical trials the choice of accrual period and the follow-up period to produce an adequate power to detect a specified difference is not unique. He presented methods for determining appropriate combinations for the accrual and follow-up periods and the unique cost effective choice of accrual and follow-up periods. It may be noted that the unique cost effective duration of follow-up depends on the ratio of the cost of accruing patients to the cost of following patients for end point determination. The optimal length of a clinical trial minimizing its cost is also considered by Gross *et al.* (1987).

Palta and Amini (1985) discuss sample size determination for survival studies, taking stratification into account. Formulas and tables, which apply to the case of multicenter clinical trials, are given. Estimation of sample size, when a covariate is grouped into strata, is also given.

For a nice review on the size and length of a phase III clinical trial, the reader is referred to George (1984).

Appendix

Guide to Selected Sample Size Tables

The percentiles of Normal, t, χ^2, and F distributions are available in many standard textbooks. The following is a partial listing of other tables that give sample sizes or those that are useful in determining the sample sizes.

Aleong, J., and Bartlett, D. E. (1979). Improved graphs for calculating sample sizes when comparing two independent binomial distributions. *Biometrics* **35,** 875–878. *Sample size charts for testing the equality of two Bernoulli probabilities with a specified power are given.*

Bechhofer, R. E. (1954). A single-sample multiple decision procedure for ranking means of normal populations with known variances. *Ann. Math. Statist.* **25,** 16–39. *Tables of λ values used in Equation (6.2.7) are given.*

Bechhofer, R. E., and Dunnett, C. W. (1988). Percentage points of multivariate student t distributions. *In Selected Tables in*

Mathematical Statistics, Vol. 11, American Mathematical Society, Providence. RI. *An extensive tabulation of* $|d|_{k-1,\,\infty}^{\alpha}$ *values used in Equation (3.2.17) and other percentiles are given.*

Bechhofer, R. E., Dunnett, C. W., and Sobel, M. (1954). A two-sample multiple decision procedure for ranking means of normal populations with a common unknown variance. *Biometrika* **41**, 170–176. *The values of h used in Section 6.2.2 are tabulated.*

Bowman, K. O., and Kastenbaum, M. A. (1975). Sample size requirement: single and double classification experiments. *In Selected Tables in Mathematical Statistics*, Vol. 3, 111–232, American Mathematical Society, Providence, RI. *Table 1 on pp. 123–194 gives the noncentrality parameter of an F-test for testing the equality of several means, and the sample sizes can be determined by using this noncentrality parameter.*

Bratcher, T. L., Moran, M. A., and Zimmer, W. J. (1970). Tables of Sample sizes in the analysis of variance. *J. Qual. Technology* **2**, 156–164. *Tables of common sample sizes, for testing the equality of several means by controlling the power, are given.*

Casgrande, J. T., Pike, M. C., and Smith, P. G. (1978). An improved approximate formula for calculating sample sizes for comparing two binomial distributions. *Biometrics* **34**, 483–486. *Exact and approximate common sample sizes are tabulated for testing the equality of two Bernoulli probabilities using a 0.05 level one-sided test with a power of 0.90.*

Cochran, W. G., and Cox, G. M. (1957). *Experimental Designs*, (Second Edition), John Wiley & Sons, New York, NY.

(i) *Table 2.1 on pp. 20–21 gives the common sample sizes for testing the equality of two population means using a* t-test *based on independent samples by controlling power.*

(ii) *Table 2.1(a) on pp. 24–25 gives the common sample sizes for testing the equality of two Bernoulli probabilities by controlling the power.*

Cohen, J. (1969). *Statistical Power Analysis for the Behavioral Sciences*, Academic Press, New York, NY. *Contains several tables giving the power functions and sample sizes for various tests discussed in the text.*

Croarkin, M. C. (1962). Graphs for determining the power of Student's *t*-test. *J. of Research of National Bureau of Standards* **B66,** 59–70. *Power charts for the t-tests are given, and these are helpful in determining the sample sizes.*

Davies, O. L. (1960). (Ed.) *The Design and Analysis of Industrial Experiments.* Oliver and Boyd, London, England.

  (i) *Table E on pp. 606–607 gives the sample size for testing one population mean using a t-test by controlling the power.*
 (ii) *Table E_1 on pp. 609–610 gives the common sample sizes for testing the equality of two population means using a t-test based on independent samples by controlling the power.*
(iii) *Table G on p. 613 gives the sample size for testing a hypothesis about a population variance.*
 (iv) *Table H on p. 614 gives the common sample size for testing the equality of two variances.*

Desu, M. M., and Sobel, M. (1968). A fixed subset-size approach to the selection problem. *Biometrika* **55,** 401–410. *An extended version of the table of λ-values of Bechhofer (1954) is given.*

Dunnett, C. W. (1955). A multiple comparisons procedure for comparing several treatments with a control. *J. Amer. Statist. Assoc.* **50,** 1096–1121.

Dunnett, C. W. (1964). New Tables for Multiple Comparisons with a Control. *Biometrics* **20,** 482–491. *Table of $|d|_{k-1,\infty}^{\alpha}$ values used in Equation (3.2.17) is given.*

Feigl, P. (1978). A graphical aid for determining sample size when comparing two independent proportions. *Biometrics* **34,** 111–122. *Sample size charts, for testing the equality of Bernoulli probabilities by controlling the power, are given.*

Gail, M., and Gart, J. J. (1973). The determination of sample sizes for use with exact conditional test in 2×2 comparative trials. *Biometrics* **29**, 441–448. *Tables of sample sizes for Fisher's exact test for differences between proportions are given.*

Gehan, E. A., and Schneiderman, M. A. (1982). Experimental design of clinical trials. *In Cancer Medicine*, Second Edition, Eds. Holland, J. F., and Frei, E., III, 531–533, Lea and Febiger, Philadelphia, PA. *Modified version of Table 2.1(a) of Cochran and Cox (1957) is given.*

George, S. L., and Desu, M. M. (1974). Planning the size and duration of a clinical trial studying the time to some critical event. *J. Chron. Dis.* **27**, 15–24.

(i) *Table 1 gives the sample size to detect a significant difference in two exponential distributions using a one-sided test and controlling the power.*

(ii) *Table 2 gives the duration of a clinical trial.*

Gibbons, J. D., Olkin, I., and Sobel, M. (1977). *Selecting and Ordering Populations: A New Statistical Methodology*, John Wiley & Sons, New York, NY.

(i) *Table Q.6 on pp. 506–507 is adapted from Desu and Sobel (1968).*

(ii) *Table G.1 on pp. 440–442 is adapted from Gupta and Sobel (1962).*

(iii) *Table R.1 on pp. 510–512 is useful in solving Equation (6.4.5).*

Gupta, S. S. (1963). On a selection and ranking procedure for gamma populations. *Ann. Inst. Statist. Math.* **14**, 199–216. *A table useful in solving Equation (6.4.5) is given.*

Gupta, S. S., and Sobel, M. (1962). On the smallest of several correlated F statistics. *Biometrika* **49**, 509–523. *A table useful in solving Equation (6.2.13) is given.*

Harter, H. L. (1960). Tables of range and studentized range. *Ann. Math. Statist.* **31**, 1122–1147. *Table of $q_{k,\infty}^{\alpha}$ used in Equation (3.2.13) is given.*

Haseman, J. K. (1978). Exact sample sizes for use with the Fisher-Irwin test of 2×2 tables. *Biometrics* **34**, 106–109. *Exact sample sizes are tabulated for testing two Bernoulli probabilities using a one-sided test controlling the power.*

Haynum, G. E., Govindarajulu, Z., and Leone, F. C. (1973). Tables of the cumulative noncentral chi-square distribution. *In Selected Tables in Mathematical Statistics*, **1**, 1–78, Ed. Harter and Owen; American Mathematical Society. *Table II on pp. 43–78 gives the noncentrality parameter of a chi-square distribution, and sample size can be determined from the noncentrality parameter.*

Korn, E. L. (1986). Sample size tables for bounding small proportions. *Biometrics* **42**, 213–216. *Sample size tables are given controlling the upper confidence limit for estimating one small Bernoulli probability.*

Lesser, M. L., and Cento, S. J. (1981). Tables of power for the *F*-test for comparing two exponential survival distributions. *J. Chron. Dis.* **34**, 533–544. *Tables of power of the two-sided F-test for comparing two exponential distribution are given, and they can be used to determine the sample size.*

Lieberman, G. J., and Owen, D. B. (1961). *Tables of the Hypergeometric Distribution*, Stanford University Press, Stanford, CA. *The probabilities and cumulative distribution of the hypergeometric distribution are tabulated.*

Odeh, R. E., and Fox, M. (1975). *Sample Size Choice: Charts for Experiments with Linear Models*, Marcel Dekker, Inc., New York, NY. *Expanded power charts for F-tests are given on pp. 83–188, and they are helpful in determining the sample sizes as discussed in the text.*

Pearson, E. S., and Hartley, H. O. (1951). Charts of the power function for analysis of variance tests, derived from the noncentral *F*-distribution. *Biometrika* **38**, 112–130. *Charts of the power function of the F-test are given.*

Rubinstein, L. V., Gail, M., and Santner, T. J. (1981). Planning the duration of a comparative clinical trial with loss to follow-up and a period of continued observation. *J. Chron. Dis.* **34,** 469–479. *Tables of T + τ of Section 7.3.2 for $\phi_i = 0$ and $\phi_i = \lambda_i$ are given.*

Schork, M. A., and Williams, G. W. (1980). Number of observations required for the comparison of two correlated proportions. *Comm. Statist.* **B9,** 349–357. *The sample sizes needed for McNemar's test by controlling power are tabulated.*

Thompson, Jr., W. A., and Endriss, J. (1961). The required sample size when estimating variances. *Amer. Statistician* **15,** 22–23. *A table of sample sizes, needed when a fixed width confidence interval for the population standard deviation is required, is given.*

Problems and Supplements

(1) Let $\hat{\theta}$ be an unbiased estimator of θ, based on a random sample of size n from a probability distribution.

(a) One wants to control the absolute error so that

$$P(|\hat{\theta} - \theta| \leq d) \geq 1 - \alpha,$$

where d and α are specified positive constants. Determine n so that this requirement is satisfied. [Hint: Use Chebychev's inequality and get an equation for determining n.]

(b) In some cases the distribution of the estimator can be approximated by a normal distribution. Using this approximation, determine n so as to meet the same specification. [Note: In these two cases, this specification for obtaining the sample size amounts to obtaining the sample size by controlling the variance of $\hat{\theta}$, and the variance bound is a function of d and α.]

(2) Suppose that T is a statistic based on a random sample of size n from a continuous probability distribution. A confidence

interval of the form $(T/f, Tf)$, with confidence coefficient $1 - \alpha$, for the positive valued parameter θ is desired, where f and α are specified constants such that $f > 1$. Assuming that the distribution of T/θ depends only on n, determine n so that

$$P\left(\frac{T}{f} < \theta < Tf\right) \geq 1 - \alpha.$$

[Note: The equation for determining n resembles (1.2.11). This confidence interval has the property

$1 -$ (lower limit/middle point) = (upper limit/middle point) $- 1$.

See Nelson (1982).]

(3) Let $X_1, X_2, \ldots, X_n$ be a random sample on $X \sim N(\mu, \sigma^2)$. Let s^2 be the sample variance. Sometimes s is used as an estimator of the population standard deviation σ. Obtain the needed equation to determine n so as to satisfy the following requirement

$$P\left(\left|\frac{s - \sigma}{\sigma}\right| \leq r^*\right) \geq 1 - \alpha,$$

where r^* and α are specified constants such that $r^* < 1$. Obtain an approximation to the required sample size by using a normal percentile approximation to the χ^2 percentile. [See Section 1.2.6 and Thompson and Endriss (1961).]

(4) *Life Testing.* Consider a life-testing experiment where n items are put on test in order to observe the life times. Let X_1, $X_2, \ldots, X_n$ be the life lengths of these items. It is assumed that the life time distribution is $NE(0, \theta)$. At times the testing of the items is stopped after observing t failures. In other words, only the first t order statistics, namely, $X_{(1)}, X_{(2)}, \ldots, X_{(t)}$ of the sample are observed. The usual estimator of θ is

$$\hat{\theta} = \frac{\sum_{i=1}^{t} X_{(i)} + (n - t)X_{(t)}}{t}.$$

Now $(2t/\theta)\hat{\theta} \sim \chi^2(2t)$. Determine the sample size n so that

$$P\left(\left|\frac{\hat{\theta} - \theta}{\theta}\right| \leq r^*\right) \geq 1 - \alpha$$

for specified constants $r^*(<1)$ and α. [Note: This specification determines a minimum value for t, which in turn determines a minimum value for n.]

Testing a hypothesis about θ: Consider the problem of testing H_0: $\theta = \theta_0$ against the one-sided alternatives H_A: $\theta > \theta_0$. The critical region of the usual α-level test is

$$\hat{\theta} > \left(\frac{\theta_0}{2t}\right) \chi^2_{1-\alpha}(2t).$$

Determine n so that this test has a power $1 - \beta$ at $\theta = \theta_1$.

(5) *Estimation by Controlling the Width of a Confidence Interval.* Let $X_1, X_2, \ldots, X_n$ be a random sample on $X \sim N(\mu, \sigma^2)$, where both the parameters are unknown.

(a) Let (L, U) be the usual confidence interval for μ with confidence coefficient $1 - \alpha$, based on this random sample. The width of this confidence interval, $U - L$, is a random variable.

  (i) Determine the sample size n so that

$$P(U - L \leqslant k\sigma) \geqslant \gamma,$$

where k and γ are given positive constants. In other words, obtain the implicit equation for determining n, which depends on the percentiles of $T(n-1)$ and $\chi^2(n-1)$.

  (ii) Since σ^2 is unknown, one may want to choose n to satisfy a modified requirement, namely,

$$P(U - L \leqslant k) \geqslant \gamma.$$

where k an γ are given positive constants. This choice of n clearly depends on σ^2. A way out of this predicament is to consider a procedure based on a sample drawn in two stages. Develop such a procedure. [See Graybill (1958).]

(b) Let W be the width of the usual confidence interval for σ^2, with confidence coefficient $1 - \alpha$. Determine n so that

$$P(W \leqslant k\sigma^2) \geqslant \gamma,$$

for specified constants k and γ. [For relevant tables see Graybill and Morrison (1960).]

(c) *Prediction Interval.* At times one may want to predict a future observation on the basis of a random sample from $N(\mu, \sigma^2)$ distribution. The usual $100(1 - \alpha)$ percent prediction interval for a future observation, when σ^2 is unknown, is

$$\bar{X} \pm t_{\alpha/2}(n - 1)\left[\left(1 + \frac{1}{n}\right)s^2\right]^{1/2},$$

where $\bar{X}$ is the sample mean, and s^2 is the sample variance. Let W_p be the width of this interval. Determine n so that

$$P(W_p \leq k\sigma) \geq \gamma,$$

for specified constants k and γ.

(6) Let Y_1 be the minimum of a sample of size n from $NE(\gamma, \theta)$ distribution. Consider $(Y_1 - d, Y_1)$ as a confidence interval for γ. Determine n such that the confidence coefficient is $1 - \alpha$, assuming that θ is known. [Desu (1971) showed that this interval is the optimal confidence interval of width d.]

(7) *Stein's Two-Stage Sampling Scheme.* An initial random sample $X_1, X_2, \ldots, X_{n_1}$ is taken on the random variable $X \sim N(\mu, \sigma^2)$. Here both the parameters μ and σ^2 are unknown and $n_1 > 3$. Let s^2 be the variance of this first stage sample and let

$$n = \max\left\{n_1, \left[\frac{s^2}{c}\right] + 1\right\},$$

where $[y]$ stands for the integral part of y. Take $n - n_1$ additional independent observations on X, if necessary. Let $\bar{X}$ be the mean of all n observations on X.

(i) Show that $\sqrt{n}(\bar{X} - \mu)/s$ is distributed as Student's t variable with $(n_1 - 1)$ degrees of freedom.

(ii) Consider the problem of estimating μ such that

$$P(|\bar{X} - \mu| \leq d) \geq 1 - \alpha,$$

where $\bar{X}$ is the mean based on the two stage sampling

scheme. To meet this requirement, show that the appropriate choice of c is $\{d/(t_{\alpha/2}(n_1 - 1))\}^2$.

(iii) If one wants to estimate μ by $\bar{X}$ so that $Var(\bar{X}) < V^*$,
show that an appropriate choice of c is $(n_1 - 3)V^*/(n_1 - 1)$.

(iv) To obtain an α-level test of H_0: $\mu = \mu_0$, based on $\bar{X}$, with
power $1 - \beta$ at $\mu_1 > \mu_0$, show that the appropriate choice
of c is

$$c = \frac{(\mu_1 - \mu_0)^2}{\{t_\alpha(n_1 - 1) + t_\beta(n_1 - 1)\}^2}.$$

[Hint: Show that the conditional distribution of the variable in (i), given s^2, is $N(0, \sigma^2/s^2)$ and then obtain the
unconditional distribution, while noting that $(n_1 - 1)s^2/\sigma^2$
is distributed as $\chi^2(n_1 - 1)$ random variable. See Stein
(1945), and Hewett and Spurrier (1983).]

(8) Let $X_1, X_2, \ldots, X_n$ be a random sample on $X \sim NE(\gamma, \theta)$.
Consider Y, the sample minimum, as an estimator of γ.

(i) Determine n, the sample size, so that $E(Y - \gamma)^2 \leqslant M^*$, a
given constant, assuming that θ is known.

(ii) When θ is unknown, one may use an estimator based on a
two-stage sampling scheme described next.

(a) Take an initial random sample $X_1, X_2, \ldots, X_{n_1}$ of size
n_1 on X, where $n_1 \geqslant 2$. Calculate

$$\hat{\theta} = \sum_{i=1}^{n_1} \frac{X_i - Y(n_1)}{(n_1 - 1)},$$

where $Y(n_1)$ is the minimum of this sample.

(b) Let $n = \max\{n_1, [c\hat{\theta}] + 1\}$.

(c) Take $n - n_1$ additional independent observations on
X, if necessary. Let $Y(n)$ be the minimum of all the n
observations.

The statistic $Y(n)$ is used as an estimator of γ. Find c such
that $E(Y(n) - \gamma)^2 \leqslant M^*$.

[Hint: Given $\hat{\theta}$, $Y(n) \sim NE(\gamma, \hat{\theta}/n)$. Also $2\hat{\theta}/\theta$ follows
$\chi^2(2n_1 - 2)$ distribution. See Desu et al. (1976).]

(9) *Estimation with a Prescribed Variance.* Let X be a random variable of interest with the probability function or probability density function $f(x, \theta)$. It is desired to estimate a function of θ, $\rho(\theta)$, with prescribed precision, that is, one wants an unbiased estimator $\hat{\rho}$ such that $\mathrm{Var}(\hat{\rho}) \leq B(\theta)$, where B is a given function. Under certain conditions, one can achieve this by taking the sample in two stages. Develop such a two-stage estimation procedure when X has a Bernoulli distribution, and one wants to estimate $\theta = E(X)$. [See Birnbaum and Healy (1960).]

(10) Let $X_1, X_2, \ldots, X_n$ be a random sample on $X \sim N(\mu, \sigma^2)$, where μ and σ^2 are unknown. Show that an approximation to the sample size required for an α-level one-sided t-test of $H_0: \mu = \mu_0$, $H_A: \mu > \mu_0$, with a power of $1 - \beta$ when μ_0 is the $100p$ percentile ($p < 0.5$) of the normal distribution with mean μ_1 (a specified value of μ under the alternative hypothesis) and variance σ^2 is

$$\left[\left\{ \frac{z_\alpha + z_\beta}{z_p} \right\}^2 + 0.5 z_\alpha^2 \right] + 1.$$

[Hint: $\bar{X} - ks$ is approximately distributed as a normal variable with mean $\mu - k\sigma$ and variance $(\sigma^2/n)(1 + k^2/2)$ for any given k value. See Guenther (1981).]

(11) In acceptance sampling using measurement data, the lot quality is usually specified by the proportion of defectives. In the case of a single upper specification limit U, any unit with a value greater than U is considered defective. The measurement variable X is assumed to follow the $N(\mu, \sigma^2)$ distribution. If p_0 is the acceptable proportion of defectives (AQL) of a lot with mean μ_0, then $U = \mu_0 + z_{p_0} \sigma$. For a lot with proportion of defectives p_1, the mean μ_1 differs from μ_0 by

$$\mu_1 - \mu_0 = \{U - z_{p_1} \sigma\} - \{U - z_{p_0} \sigma\} \equiv \Delta \sigma.$$

Show that the power function of the t-test, considered in Problem 10, for testing the null hypothesis that the proportion of defectives is p_0 against the alternative that it is greater than p_0 is

$$P(T(n - 1; (z_{p_0} - z_p)\sqrt{n}) > t_\alpha(n - 1)).$$

Show that the sample size, required for the power to be $1 - \beta$ at

$p = p_1$, satisfies the following implicit relationship

$$t_{1-\beta}(n - 1; \Delta\sqrt{n}) = t_\alpha(n - 1).$$

[See Chapter 2 of Guenther (1977).]

(12) Let $X_1, X_2, \ldots, X_n$ be a random sample on X, which is uniformly distributed over $[0, \theta]$ for a positive θ. An unbiased estimator of θ is $\hat{\theta} = (n + 1)X_{[n]}/n$, where $X_{[n]}$ is the largest observation. Determine an implicit equation for the sample size n such that $P[\hat{\theta}/\theta > 1 - r^*] = 1 - \alpha$.

(13) *Sample-size Precision Schedule.* In the problem of estimation, the required sample size depends on d or r^*, for a given α. As such the quantity d or r^* may be viewed as a measure of the precision of the estimator. Since the cost of the experiment depends to a great extent on the sample size, the experimenter may want at times to tabulate d or r^* obtainable for various values of n and then make a choice. Such a table is the so-called "sample-size precision schedule" (*cf.* Mace, 1964). Consider the problem of estimating $\theta = E(X)$, where $X \sim NE(0, \theta)$. The usual estimator is $\bar{X}$ (mean of a sample of size n), and this estimator has to satisfy the condition

$$P\left(\left|\left(\frac{\bar{X}}{\theta}\right) - 1\right| \leqslant r^*\right) = 0.9.$$

Prepare a sample-size precision schedule for $n = 2\,(1)\,10\,(10)\,50$. [Hint:

$$r^* = \frac{\chi_{0.95}^2(2n) - \chi_{0.05}^2(2n)}{\chi_{0.95}^2(2n) - \chi_{0.05}^2(2n)}.$$

(14) Let X be a random variable with $N(\mu, \sigma^2)$ distribution, where the variance σ^2 is known. The prior distribution of μ is $N(\mu_0, \sigma_0^2)$ distribution. Then the posterior distribution of μ, given a random sample X of size n on X, is $N(\hat{\mu}, \sigma_1^2)$, where

$$\hat{\mu} = \frac{\{(\mu_0/\sigma_0^2) + (n\bar{X}/\sigma^2)\}}{(1/\sigma_0^2) + (n/\sigma^2)}$$

and

$$\sigma_1^2 = \{(1/\sigma_0^2) + (n/\sigma^2)\}^{-1};$$

here $\bar{X}$ is the sample mean. The Bayes (point) estimate of μ is $\hat{\mu}$. Determine n so that $\hat{\mu} \pm d$ is a posterior interval of level $1 - \alpha$. In other words determine n such that

$$P(|\mu - \hat{\mu}| \le d \,|\, \underline{X}) \ge 1 - \alpha,$$

for given positive d and α.

(15) Let X be a random variable with $N(\mu, \sigma^2)$ distribution, where both parameters are unknown. Consider the uniform prior, namely,

$$p(\mu, \sigma^2) \propto \left(\frac{1}{\sigma^2}\right).$$

Then the (marginal) posterior distribution of σ^2, given a random sample $\underline{X}$ of size $n(>3)$ on X, is such that $(n-1)s^2/\sigma^2 \sim \chi^2(n-1)$, where s^2 is the sample variance. The Bayes estimate of σ^2 is

$$\tilde{\sigma}^2 = \frac{(n-1)s^2}{n-3}.$$

Determine n so that

$$P\left(\left|\frac{\sigma^2 - \tilde{\sigma}^2}{\tilde{\sigma}^2}\right| \le r \,\Big|\, \underline{X}\right) \ge 1 - \alpha,$$

for given positive fractions r and α. [Note: The equation to determine n resembles (1.2.11) and as such an approximation can be derived using the discussion of that section.]

(16) Let X be random variable with $NE(0, 1/\lambda)$ distribution. Consider the uniform prior for λ, namely, $p(\lambda) \propto k$. The posterior density function, given a random sample $\underline{X}$ of size n on X, is

$$\pi(\lambda \,|\, \underline{X}) = t^{n+1}\lambda^n e^{-t\lambda}/(n!), \qquad \text{for } \lambda > 0,$$

where t is the sample sum. The Bayes estimate is $\hat{\lambda} = (n+1)/t$. Determine n so that

$$P\left(\frac{|\lambda - \hat{\lambda}|}{\hat{\lambda}} \le r \,\Big|\, \underline{X}\right) \ge 1 - \alpha,$$

for given positive fractions r and α.

(17) Let X and Y be two independent random variables such that $X \sim N(\mu_1, \sigma^2)$ and $Y \sim N(\mu_2, \sigma^2)$. First-stage samples $X_1, \ldots, X_{n_1}$ on X and $Y_1, \ldots, Y_{n_1}$ on Y are available. Let s_p^2 be the pooled variance of the two samples and let

$$n = \max\left\{ n_1, \left[\frac{s_p^2}{c} \right] + 1 \right\}.$$

An additional sample of $n - n_1$ observations on X and Y are obtained, if necessary. The mean of all n observations on X is $\bar{X}$, and the mean of all n observations on Y is $\bar{Y}$.

(i) Show that $\sqrt{n}[(\bar{X} - \bar{Y}) - (\mu_1 - \mu_2)]/s_p$ has Student's t-distribution with $2(n_1 - 1)$ degrees of freedom.

(ii) In the problem of estimating $\mu_1 - \mu_2$, show that the choice of c, to meet the requirement $P(|(\bar{X} - \bar{Y}) - (\mu_1 - \mu_2)| < d^*) = 1 - \alpha$, is

$$c = \frac{1}{2} \left\{ \frac{d^*}{t_{\alpha/2}(2n_1 - 2)} \right\}^2$$

(iii) Show that the choice of c in relation to the problem of testing $H_0: \mu_1 = \mu_2$ against the alternative $H_A: \mu_1 > \mu_2$ is

$$c = \frac{(\mu_1 - \mu_2)^2}{2}(t_\alpha(\nu) + t_\beta(\nu))^{-2}$$

where $\nu = 2n_1 - 2$.

(18) Let $X_1, X_2, \ldots, X_n$ be a random sample on $X \sim NE(\gamma_1, \theta)$ and $Y_1, Y_2, \ldots, Y_n$ be an independent random sample on $Y \sim NE(\gamma_2, \theta)$. Let T_1 and T_2 be the minimums of the two samples respectively. If θ is known, show that the common sample size n such that $E\{(T_1 - T_2) - (\gamma_1 - \gamma_2)\}^2 \leq M^*$ is $[\theta\sqrt{2/M^*}] + 1$. Design a two-stage estimation procedure when θ is unknown. [See Desu et al. (1976).]

(19) Let $X_1, X_2, \ldots, X_n$ be a random sample on $X \sim N(\mu_1, \sigma^2)$ and $Y_1, Y_2, \ldots, Y_n$ be a sample on $Y \sim N(\mu_2, \sigma^2)$, where μ_1, μ_2, and σ^2 are unknown. Let x_{p_1} and y_{p_2} be $100p_1$ and $100p_2$ percentiles of X and Y variables, respectively. For an α-level test of

$$H_0: x_{p_1} = y_{p_2}, \quad \text{against } H_A: x_{p_1} < y_{p_2}, \quad (p_1 > p_2)$$

one wishes to use the usual one-sided two sample t-test based on the statistic

$$T = \frac{(\bar{X} - \bar{Y})}{s_p \sqrt{\dfrac{2}{n}}},$$

which follows a noncentral t-distribution with $2n - 2$ degrees of freedom and noncentrality parameter $\sqrt{n/2}\ (z_{p_2} - z_{p_1})$, when H_0 is true. If a power of $1 - \beta$ is desired when $x_{p_1} = y_{p'_2}$, where $p'_2 < p_2$, an approximation to the required sample size n is $[n^*] - 1$, where

$$n^* = (2 + k^2)\left(\frac{z_\alpha + z_\beta}{z_{p'_2} - z_{p_2}}\right)^2;$$

here the constant k is given by

$$2k = \frac{z_\alpha(z_{p'_2} - z_{p_1}) + z_\beta(z_{p_2} - z_{p_1})}{z_\alpha + z_\beta}.$$

[See Guenther (1975).]

(20) Let $X_1, X_2, \ldots, X_n$ be a random sample on $X \sim N(\mu_1, \sigma_1^2)$ and let $Y_1, Y_2, \ldots, Y_n$ be a random sample on $Y \sim N(\mu_2, \sigma_2^2)$. Assume that X and Y are independent. Let $R = P(Y < X)$. In stress and strength analysis, X is the strength of a component subject to a stress Y so that the resulting reliability of the component is R. The samples are used to estimate R. Let $\bar{X}$ and $\bar{Y}$ be the two-sample means and let s_x^2 and s_y^2 be the two sample variances. An estimate of R is $\hat{\delta} = (\bar{X} - \bar{Y})/s$, where $s^2 = s_x^2 + s_y^2$. Given R_1, R_2, α and β find n and δ_c such that

$$P(\hat{\delta} > \delta_c \mid \delta = \delta_1) = 1 - \alpha,$$

$$P(\hat{\delta} > \delta_c \mid \delta = \delta_2) = \beta,$$

where $\delta_i = \Phi^{-1}(R_i)$, $i = 1, 2$, Φ being the distribution function of the standard normal distribution. [Hint: The probabilities of interest depend on noncentral t-distributions, and the equations have to be solved iteratively. For an approximate solution, see Reiser and Guttman (1989).]

(21) Let $X_1, X_2, \ldots, X_k$ be k independent random variables such that $X_i \sim N(\mu_i, \sigma^2)$, $i = 1, \ldots, k$. Random samples of size n_1 are taken on each X_i. The pooled sample variance is denoted by s_p^2. It is well known that $k(n_1 - 1). s_p^2/\sigma^2 \equiv vs_p^2/\sigma^2$ follows the $\chi^2(v)$ distribution. Let

$$n = \max\left\{n_1, \left[\frac{s_p^2}{c}\right] + 1\right\},$$

where c is a specified constant. Now, if necessary, random samples of size $n - n_1$ are taken on each X_i and the mean of the n observations on X_i is $\bar{X}_i$, $i = 1, \ldots, k$. Let $T_i = \sqrt{n}(\bar{X}_i - \mu_i)/s_p$, $i = 1, \ldots, k$. The distribution of the random vector $T = (T_1, T_2, \ldots, T_k)'$ is a k-variate t-distribution of Dunnett and Sobel (1955). If one wants to control the error in estimating the differences $\mu_i - \mu_j$ as in Equation (3.2.10), show that the choice of c is

$$c = (d/q_{k,v}^\alpha)^2,$$

where $q_{k,v}^\alpha$ is the upper α-percentile of the studentized range distribution with parameters k and v as tabulated in Harter (1959). With this choice of c, a $100(1 - \alpha)$ confidence region for the differences is

$$\mu_i - \mu_j \in (X_i - X_j) \pm q_{k,v}^\alpha \left(\frac{s_p}{\sqrt{n}}\right).$$

(22) Following the notation of Section 3.2.2 and Problem 17, if one wants to control the error in estimating the $k - 1$ differences $\mu_i - \mu_1$ for $i = 2, 3, \ldots, k$ as specified in Equation (3.2.15), show that the choice of c for an appropriate two-stage procedure is

$$c = \frac{d^2}{2h^2(v; 0.5)},$$

where $h(v; \rho)$ are the percentiles of $(k - 1)$ variate t-distribution tabulated by Bechhofer and Dunnett (1988). [See Desu (1989).]

(23) Let $X_i \sim N(\mu_i, \sigma^2)$, $i = 1, 2, \ldots, k$ be k independent random variables. Samples of size n are taken on each X_i and let s_i^2 be the variance of the sample on X_i. Testing the null hypothesis $H_0: \sigma_1^2 = \sigma_2^2 = \cdots = \sigma_k^2$ is equivalent to testing the null hypothesis

H_0: $\ln \sigma_1^2 = \ln \sigma_2^2 = \cdots = \ln \sigma_k^2$. Note that $\ln s_i^2$ is approximately distributed as $N(\ln \sigma_i^2, 2/n)$, $i = 1, 2, \ldots, k$. Thus, H_0 can be tested using the test statistic

$$T = \frac{\sum\limits_{i=1}^{k} (\ln s_i^2 - A)^2}{k - 1},$$

where

$$A = \Sigma_{i=1}^{k} \ln s_i^2 / k.$$

(i) Find the critical region of an α-level test of H_0 using T when the alternative is not H_0.

(ii) Find the required sample size controlling the power of this test at $1 - \beta$, when $\sigma_i^2 = \sigma_i^{2*}$ for specified σ_i^{2*}, $i = 1, 2, \ldots, k$.

(24) Warner (1965) developed an ingeneous method for estimating the proportion, θ, of individuals belonging to a sensitive category C. In its simplest form it consists of taking a simple random sample of size n with replacement. Each individual in the sample is requested to answer question (a) or (b) with a "yes" or "no" response. The individual selects the question (a) with probability p. The two questions are

(a) I belong to the sensitive category C;
(b) I do not belong to the sensitive category C.

The individual will not disclose which question is being chosen. If n_1 individuals gave a "yes" response, then an estimate of θ is $\hat{\theta} = \{(n_1/n) - (1 - p)\}/(2p - 1)$, and the variance of $\hat{\theta}$ is $\{\theta(1 - \theta)/n\} + \{p(1 - p)/(n(2p - 1)^2)\}$. Suppose that $\hat{\theta}_2$ is the usual estimate of θ based on a simple random sample of size n^*, where the sample is drawn with replacement and the technique is not used. Show that for the two estimates to have equal variances, the sample size n has to be taken as

$$n = n^* \left\{ 1 + \frac{p(1 - p)}{(2p - 1)^2 \theta(1 - \theta)} \right\}.$$

(25) Let $\pi_1, \pi_2, \ldots, \pi_k$ be k populations where observations from π_i follow $NE(\mu_i, \theta)$ distribution, $i = 1, 2, \ldots, k$. It is of interest to select the best population, the one associated with the largest μ_i. The usual selection procedure is based on Y_1, $Y_2, \ldots, Y_k$ where Y_i is the minimum of a sample of size n from π_i, $i = 1, 2, \ldots, k$. The population that corresponds to max Y_i is selected as the best population. The problem is to determine the common sample size n so that the $P(cs) \geq P^*$ whenever $\mu_{[k]} - \mu_{[k-1]} \geq \delta^*$. ($\mu_{[1]} \leq \mu_{[2]} \leq \cdots \leq \mu_{[k]}$ are the ordered μ values). Show that the required n value is to be chosen so that

$$n\left(\frac{\delta^*}{\theta}\right) \geq -\ln \nu$$

where ν is the solution of

$$(\nu k)^{-1}\{1 - (1 - \nu)^k\} = P^*.$$

[See Raghavachari and Starr (1970).]

(26) In the selection problem of (25), the common sample size depends on θ. When θ is unknown, in order to meet the requirement on $P(cs)$ one needs to use a two-stage procedure similar to the one described in Section 6.2.2. Design such a two-stage procedure. [See Desu *et al.* (1977).]

(27) Let $X_{ij1}, \ldots, X_{ijn}$ be a random sample of size n on $X_{ij} \sim N(\mu_{ij}, \sigma^2)$. It is known that

$$\mu_{ij} = \mu + \theta_i + \beta_j + \gamma_{ij},$$

for $i = 1, 2, \ldots, k$ and $j = 1, 2, \ldots, b$. It is assumed that $\Sigma_i \theta_i = 0$, $\Sigma_j \beta_j = 0$, $\Sigma_i \gamma_{ij} = 0$ for $j = 1, 2, \ldots, b$, and $\Sigma_j \gamma_{ij} = 0$ for $i = 1, 2, \ldots, k$. It is of interest to test the null hypothesis $H_0 \colon \theta_1 = \theta_2 = \cdots = \theta_k$. The problem is to determine n so that the usual α-level F-test for testing H_0 has a power not less than $1 - \beta$ for certain alternatives. Obtain an expression for the power function of the usual F-test and indicate how one can determine n. [See Kastenbaum *et al.* (1970).]

(28) Let (X_i, Y_i), $i = 1, 2, \ldots, n$ be a random sample from a bivariate normal distribution with mean vector $(\mu_x, \mu_y)'$ and

covariance matrix

$$\begin{bmatrix} \sigma_x^2 & \rho\sigma_x\sigma_y \\ \rho\sigma_x\sigma_y & \sigma_y^2 \end{bmatrix}.$$

Let $\beta = \rho\sigma_y/\sigma_x$. Further, let $\bar{X} = \Sigma X_i/n, \bar{Y} = \Sigma Y_i/n, \Sigma x_i^2 = \Sigma(X_i - \bar{X})^2$, $\Sigma x_i y_i = \Sigma(X_i - \bar{X})(Y_i - \bar{Y})$, $\hat{\beta} = \Sigma x_i y_i/\Sigma x_i^2$. Then Thigpen and Paulson (1974) showed that

$$T = (\hat{\beta} - \beta)/[\sigma_y^2(1 - \rho^2)/\{(n - 1)\sigma_x^2\}]^{1/2},$$

has Student's t-distribution with $n - 1$ d.f. Show that the required sample size n controlling relative error in estimating β at δ with probability $1 - \gamma$ is $[n^*] + 1$, where n^* is the solution of

$$n = \left\{ \frac{t_{\gamma/2}(n - 1)\sqrt{(1 - \rho^2)/\rho^2}}{\delta} \right\}^2 + 1.$$

[*cf.* Thigpen (1987).]

(29) Consider a study that is undertaken to obtain an upper bound on θ the probability of exhibiting some side effects when a drug is given. Determine the sample size (the number of patients to be enrolled in the study) so that the study will be successful with a 90% probability. A study will be considered successful if the upper 95% confidence limit on θ is 5% or less. [Hint: See Korn (1986).]

(30) Following the notation of Section 3.2, in clinical trials context one will be interested in testing the null hypothesis

$$H_0: \mu_i - \mu_1 = 0, \qquad i = 2, 3, \ldots, k,$$

against the alternative

$$H_A: \mu_i - \mu_1 \neq 0 \text{ for at least one } i = 2, 3, \ldots, k.$$

The critical region of the usual test, assuming equal known σ^2 for the k populations and equal sample sizes n, is

$$\max_{i=2,3,\ldots,k} |\bar{X}_i - \bar{X}_1| > c\sqrt{\frac{2\sigma^2}{n}},$$

where c will be chosen so that the test has significance level α. It may be desirable that the test has a specified power $1 - \beta$, when at least one experimental treatment i for $i = 2, 3, \ldots, k$ is sufficiently different from the control treatment one, that is,

$$\max_{i=2,3,\ldots,k} |\Delta_i| \geq d^*,$$

where $\Delta_i = \mu_i - \mu_1$ and $d^*(>0)$ is specified. Develop a procedure to determine the sample size in this context [see Bristol (1989)].

References

ALAM, K., and RIZVI, M. H. (1967). On non-central chi-squared and non-central F distributions. *Amer. Statistician* **21**, 21–22.

ANDERSON, S., and HAUCK, W. W. (1983). A new procedure for testing equivalence in comparative bioavailability and other clinical trials. *Commun. in Statist.* **A12**, 2663–2692.

BECHHOFER, R. E. (1954). A single-sample multiple decision procedure for ranking means of normal populations with known variances. *Ann. Math. Statist.* **25**, 16–39.

BECHHOFER, R. E., and DUNNETT, C. W. (1988). Percentage points of multivariate student *t* distributions. *In Selected Tables in Mathematical Statistics,* Vol. 11, American Mathematical Society, Providence, RI.

BECHHOFER, R. E., DUNNETT, C. W., and SOBEL, M. (1954). A two-sample multiple decision procedure 'for ranking means of normal populations with a common unknown variance. *Biometrika* **41**, 170–176.

BIRNBAUM, A., and HEALY, W. C., Jr. (1960). Estimates with prescribed variance based on two-stage sampling. *Ann. Math. Statist.* **31**, 662–676.

125

BOWMAN, K. O., and KASTENBAUM, M. A. (1975). Sample size requirement: single and double classification experiments. *In Selected Tables in Mathematical Statistics*, Vol. 3, 111–232 (Edited by Institute of Mathematical Statistics), American Mathematical Society, Providence, RI.

BRISTOL, D. R. (1989). Designing clinical trials for two-sided multiple comparison with a control. *Controlled Clinical Trials* **10**, 142–152.

BRISTOL, D. R. (1989). Sample sizes for constructing confidence intervals and testing of hypotheses. *Statist. Medicine* **8**, 803–811.

BRUSH, G. G. (1988). *How to Choose the Proper Sample Size*. ASQC Statistical How-To Series, Vol 12, ASQC Quality Press, Milwaukee, WI.

CASE, L. D., MORGAN, T. M., and DAVIES, C. E. (1987). Optimal restricted two-stage designs. *Controlled Clinical Trials* **8**, 146–156.

CHAPMAN, D. G. (1950). Some two-sample tests. *Ann. Math. Statist.* **21**, 601–606.

COCHRAN, W. G. (1954). Some methods for strengthening the common χ^2 tests. *Biometrics* **10**, 417–451.

COCHRAN, W. G. (1977). *Sampling Techniques* (Third Edition), John Wiley and Sons, New York, NY.

COCHRAN, W. G., and COX, G. M. (1957). *Experimental Designs* (Second Edition), John Wiley and Sons, New York, NY.

COHEN, J. (1987). *Statistical Power Analysis* (Second Edition), Lawrence Erlbaum Associates, Inc., Hillsdale, NJ.

CONNOR, R. J. (1987). Sample size for testing differences in proportions for the paired sample design. *Biometrics* **43**, 207–212.

COX, D. R. (1952). Estimation by double sampling. *Biometrika* **39**, 217–227.

DEGROOT, M. H. (1970). *Optimal Statistical Decisions*, McGraw-Hill Book Co., New York, NY.

DESU, M. M. (1971). Optimal confidence intervals of fixed width. *Amer. Statistician* **25**, 27–29.

DESU, M. M. (1989). Planning of experiments for comparing treatments with a control. Paper presented at the Annual Meetings of the American Statistical Association.

DESU, M. M., and SOBEL, M. (1968). A fixed subset-size approach to the selection problem. *Biometrika* **55**, 401–410.

DESU, M. M., NARULA, S. C., and VILLARREAL, B. (1976). Sample size determination for estimation of guarantee times. *Appl. Statist.* **25**, 275–279.

DESU, M. M., NARULA, S. C., and VILLARREAL, B. (1977). A two-stage procedure for selecting the best of k exponential distributions. *Commun. in Statist.* **A6**, 1223–1230.

DUNNETT, C. W., and GENT, M. (1977). Significance testing to establish equivalence between treatments, with special reference to data in the form of 2×2 tables. *Biometrics* **33**, 593–602.

DUNNETT, C. W., and SOBEL, M. (1955). Approximation to the probability integral and certain percentage points of a multivariate analogue of student's t-distribution. *Biometrika* **42**, 256–260.

FREEDMAN, L. S. (1982). Tables of the number of patients required in clinical trials using the Logrank test. *Statist. in Medicine* **1**, 121–129.

GAIL, M., and GART, J. J. (1973). The determination of sample sizes for use with exact conditional test in 2×2 comparative trials. *Biometrics* **29**, 441–448.

GEORGE, S. L. (1984). The required size and length of a phase III clinical trial. *In Cancer Clinical Trials: Methods and Practice,* (Edited by M. E. Buyse, M. J. Staquet, and R. J. Sylvester), Oxford University Press, New York, NY.

GEORGE, S. L., and DESU, M. M. (1974). Planning the size and duration of a clinical trial studying the time to some critical event. *J. Chron. Dis.* **27**, 15–29.

GOLDMAN, A. (1963). Sample size for a specified width confidence interval on the ratio of variances from two independent normal populations. *Biometrics* **19**, 465–477.

GOLDSTEIN, R. (1989). Power and sample size via MS/PC-DOS computers. *Amer. Statistician* **43**, 253–260.

GRAYBILL, F. A. (1958). Determining sample size for a specified width confidence interval. *Ann. Math. Statist.* **29**, 282–287.

GRAYBILL, F. A., and MORRISON, R. D. (1960). Sample size for a specified width confidence interval on the variance of a normal distribution. *Biometrics* **16**, 636–641.

GROSS, A. J., HUNT, H. H., CANTOR, A. B., and CLARK, B. C. (1987). Sample size determination in clinical trials with emphasis on exponentially distributed responses. *Biometrics* **43**, 875–883.

GUENTHER, W. C. (1975). A sample size formula for a non-central t-test. *Amer. Statistician* **29**, 120–121.

GUENTHER, W. C. (1977). *Sampling Inspection in Statistical Quality Control,* Macmillan Publishing Co., New York, NY.

GUENTHER, W. C. (1979). The use of noncentral F approximations for calculating power and sample size. *Amer. Statistician* **33**, 209–210.

GUENTHER, W. C. (1981). Sample size formulas for normal theory *t*-tests. *Amer. Statistician* **35**, 243–244.

GUPTA, S. S., and SOBEL, M. (1962). On the smallest of several correlated F statistics. *Biometrika* **49**, 509–523.

GUTTMAN, I., WILKS, S. S., and HUNTER, J. S. (1982) *Introductory Engineering Statistics* (Third Edition), John Wiley and Sons, New York, NY.

HALPERIN, M., ROGOT, E., GURIAN, J., and EDERER, F. (1968). Sample sizes for medical trials with special reference to long-term therapy. *J. Chron. Dis.* **21**, 13–24.

HARTER, H. L. (1960). Tables of range and studentized range. *Ann. Math. Statist.* **31**, 1122–1147.

HARTER, H. L. (1969). Order Statistics and Their Use in Testing and Estimation. Vol. 1: Tests Based on Range and Studentized Range of Samples from a Normal Population. Aerospace Research Laboratories, Office of Aerospace Research, U.S. Air Force.

HASEMAN, J. K. (1978). Exact sample sizes for use with the Fisher–Irwin test of 2×2 tables. *Biometrics* **34**, 106–109.

HAYNUM, G. E., GOVINDARAJULU, Z., and LEONE, F. C. (1973). Tables of the cumulative non-central Chi-square distribution. *In Selected Tables in Mathematical Statistics,* Vol. 1, 1–78 (Edited by the Institute of Mathematical Statistics), American Mathematical Society, Providence, R.I.

HEWETT, J. E., and SPURRIER, J. D. (1983). A survey of two-stage tests of hypotheses: theory and application. *Commun. in Statist.* **A12**, 2307–2425.

HOCHBERG, Y., and LACHENBRUCH, P. A. (1976). Two-stage multiple comparison procedures based on the studentized range. *Commun. in Statist.* **A5**, 1447–1453.

KASTENBAUM, M. A., HOEL, D. G., and BOWMAN, K. O. (1970) Sample size requirements: randomized block designs. *Biometrika* **57**, 573–577.

KOOPMANS, L., and QUALLS, C. (1971). Fixed length confidence intervals for parameters of the normal distribution based on two-stage sampling procedures. *Rocky Mountain J. Math.* **1**, 587–602.

KORN, E. L. (1986). Sample size tables for bounding small proportions. *Biometrics* **42**, 213–216.

LACHIN, J. M. (1977). Sample size determination for $r \times c$ comparative trials. *Biometrics* **33**, 315–324.

LACHIN, J. M. (1981). Introduction to sample size determination and power analysis for clinical trials. *Controlled Clin. Trials* **2**, 93–113.

LAKATOS, E. (1986). Sample sizes for clinical trials with time dependent rates of losses and noncompliance. *Controlled Clin. Trials* **7**, 189–199.

LEE, M. L., KRANTROWITZ, J. L., and MULLIS, C. E. (1984). Sample size, cross over trials and the problem of treatment equivalence. *Proceedings of the Biopharmaceutical Section. Amer. Statist. Assoc.* 78–80.

LIEBERMAN, G. J. and OWEN, D. B. (1961). *Tables of the Hypergeometric Distribution.* Stanford University Press, Stanford, CA.

MACE, A. E., (1964). *Sample Size Determination.* Reinhold Publishing Co., New York. (Reprinted in 1973, by R. E. Krieger Publishing Co. New York.)

MAKUCH, R. W., and SIMON, R. M. (1982). Sample size requirements for comparing time-to-failure among k treatment groups. *J. Chron. Dis.* **35**, 861–867.

McHUGH, R. B. (1961). Confidence interval inference and sample size determination. *Amer. Statistician* **15**, 14–17.

MORGAN, T. M., (1985). Planning the duration of accrual and follow-up for clinical trials. *J. Chron. Dis.* **12**, 1009–1018.

MOSHMAN, J. (1958). A method for selecting the size of the initial sample size in the Stein's two-sample procedure. *Ann. Math. Statist.* **29**, 1271–1275.

NELSON, W. (1982). *Applied Life Data Analysis,* John Wiley & Sons, New York, NY.

NOETHER, G. E. (1987). Sample size determination for some common nonparametric tests. *J. Amer. Statist. Assoc.* **82**, 645–647.

ODEH, R. E., and FOX, M. (1975). *Sample Size Choice: Charts for Experiments with Linear Models,* Marcel Dekker, Inc., New York, NY.

ODEH, R. E., CHOU, Y.-M., and OWEN, D. B. (1987). The precision for coverages and sample size requirements for normal tolerance intervals. *Commun. in Statist.* **B16**, 969–985.

ODEH, R. E., CHOU, Y.-M., and OWEN D. B. (1989). Sample-size Determination for two-Sided β-expectation tolerence intervals for a normal distribution. *Technometrics.* **31**, 461–468.

PALTA, M., and AMINI, S. B. (1985). Consideration of covariates and stratification in sample size determination for survival time studies. *J. Chron. Dis.* **38**, 801–809.

PEARSON, E. S., and HARTLEY, H. O. (1951). Charts of the power function for analysis of variance tests, derived from the noncentral *F*-distribution. *Biometrika* **38**, 112–130.

RAGHAVACHARI, M., and STARR, N. (1970). Selection problems for some terminal distributions. *Metron* **28**, 185–197.

RAO, B. R. (1986). Joint distribution of simultaneous exposures to several carcinogens in a case-control study: sample size determination. *Commun. in Statist.* **A15**, 3035–3065.

REISER, B., and GUTTMAN, I. (1989). Sample size choice for reliability verification in strength stress models. *Canad. J. Statist.* **17**, 253–259.

RODARY, C., COMNOUGUE, C., and TOURNADE, M.-F. (1989). How to establish equivalence between treatments: a one-sided clinical trial in pediatric oncology. *Statist. in Medicine* **8**, 593–598.

RUBINSTEIN, L. V., GAIL, M. H., and SANTNER, T. J. (1981). Planning the duration of a comparative clinical trial with loss to follow-up and a period of continued observation. *J. Chron. Dis.* **34**, 469–479.

SADOOGHI-ALVANDI, M. (1986). The choice of subsample size in two-stage sampling. *J. Amer. Statist. Assoc.* **81**, 555–558.

SCHOENFELD, D. A. (1981). The asymptotic properties of nonparametric tests for comparing survival distributions. *Biometrika* **65**, 316–319.

SCHOENFELD, D. A., and RICHTER, J. R. (1982). Nomograms for calculating the number of patients needed for a clinical trial with survival as an endpoint. *Biometrics* **38**, 163–170.

SCHORK, M. A., and WILLIAMS, G. W. (1980). Number of observations required for the comparison of two correlated proportions. *Commun. in Statist.* **B9**, 349–357.

SCHWERTMAN, N. C. (1987). An alternative procedure for determining analysis of variance sample size. *Commun. in Statist.* **B16**, 957–968.

SEELBINDER, B. M. (1953). On Stein's two-stage procedure. *Ann. Math. Statist.* **24**, 640–649.

STEIN, C. (1945). A two-sample test for a linear hypothesis whose power is independent of variance. *Ann. Math. Statist.* **16**, 243–258.

THALL, P. F., SIMON, R., ELLENBURG, S. E., and SHRAEGER, R. (1988). Optimal two-stage designs for clinical trials with binary response. *Statist. in Medicine.* **7**, 571–579.

THALL, P. F., SIMON, R., and ELLENBURG, S. E. (1988). Two-stage selection and testing designs for comparative clinical trials. *Biometrika* **75,** 303–310.

THIGPEN, C. C. (1987). A sample size problem in simple linear regression. *Amer. Statistician* **41,** 214–215.

THIGPEN, C. C., and PAULSON, A. S. (1974). A multiple range test for analysis of covariance. *Biometrika* **61,** 479–484.

THOMPSON, W. A., Jr., and ENDRISS, J. (1961). The required sample size when estimating variances. *Amer. Statistician* **15,** 22–23.

WARNER, S. L. (1965). Randomized response: a survey technique for eliminating evasive answer bias. *J. Amer. Statist. Assoc.* **60,** 63–69.

WOOLSON, R. F., BEAN, J. A., and ROJAS, P. B. (1986). Sample size for case-control studies using Cochran's statistic. *Biometrics* **42,** 927–932.

WORMLEIGHTON, R. (1960). A useful generalization of the Stein two-sample procedure. *Ann. Math. Statist.* **31,** 217–221.

WU, M., FISHER, M., and DeMETS, D. (1980). Sample sizes for long-term medical trials with time-dependent dropout and event rates. *J. Chron. Dis.* **1,** 109–121.

Index

A

Acceptance Sampling, 66, 114
Accuracy, 1, 4
Asymptotically, 6, 13, 15, 28, 93

B

Bayes estimate, 116
Best
  Bernoulli, 78, 80
  exponential, 80
  normal, 74, 76
  population, 73
Bioequivalence, 88, 95

C

Cauchy–Scwartz inequality, 29, 42, 68, 71
Central limit theorem, 6, 13
Confidence ellipsoid, 43

Confidence interval, 110, 111
Control, 41, 85, 91, 123
Correct selection, *see also* CS, 74, 77
Cost, total, 71
  expected, 72
Cost function, 26, 29, 38, 42, 68, 69
Crossover design, 84, 97
CS, *see also* Correct selection, 74, 75, 77, 78, 79, 81, 121

D

$|d|_{k-1}^{\alpha}$, 45
Distribution
  Bernoulli, 12, 36, 49, 78, 79
  binomial, 13
  chi-square, 2, 81
  exponential, 16, 27, 51, 80
  F, 27
  hypergeometric, 66
  k-variate t, 119

133

Distribution (*continued*)
  noncentral χ^2, 50, 52, 58, 59, 87, 88, 89, 90
  noncentral F, 46, 60
  noncentral t, 56, 95, 97, 118
  normal, 2, 23, 46
  survival, 99

E

Error
  absolute, 3, 13, 17, 24, 44, 45, 46, 55, 109
  relative, 5, 18, 27, 37, 55, 56, 61, 122
Estimation
  γ (exponential location), 17
  θ (exponential scale), 18
  θ (probability of success), 13
  μ (normal mean), 3, 112
  $\mu_1 - \mu_2$ (difference between normal means), 24, 24, 117
  σ^2 (normal variance), 5
  Bayesian approach, 20
  difference of means, 41, 42
  variance ratio, 27
Estimator, 3, 5, 17, 18, 24, 25, 27, 42, 43, 49, 52, 68, 71, 109, 110, 113, 114, 115
Expected cost, 71

G

Generalized variance, 63, 64

I

Incomplete beta function, 15
Interval
  confidence, 110, 111
  posterior, 21
  prediction, 112
  tolerance, 12

L

Least favorable configuration, 36
Life testing, 110

N

Noncentrality parameter, 46, 47, 50, 52, 59, 87, 90, 95

P

p-value, 96
Population
  best, 73, 121
  finite, 63
  mean, 64, 67, 69
  proportion, 65
  variance, 64, 67
Posterior interval, 21
Power function, 48, 56, 57, 58, 60, 61
Precision schedule, 115
Prediction interval, 21
Proportional allocation, 68

Q

$q_{k,\infty}^{\alpha}$, 45

R

Response variable
  qualitative, 83, 88
  quantitative, 83, 94

S

Sample size
  for cluster sampling, 72
  for double sampling, 70
  for estimating parameters, 3, 6, 14, 17, 18, 19, 21, 24, 28, 29, 38, 42, 43, 44, 45, 46, 64, 65, 68, 69

for selecting best population, 75, 78, 80, 81

for testing hypothesis, 7, 9, 11, 12, 16, 19, 30, 31, 35, 37, 39, 47, 49, 50, 53, 66, 86, 88, 97, 99

Sampling
acceptance, 66, 114
double, 69
simple random, 63, 120
stratified random, 67
two-stage, 49, 112, 113
two-stage cluster, 70

Study
case-control, 91
crossover, 97

T

Test
chi-square, 55
F, 49, 55, 61, 121
Fisher's exact, 37

McNemar, 84, 87
one-sided, 7, 10, 11, 12, 16, 19, 20, 30, 31, 33, 35, 37, 39
t, 55
two-sided, 9, 30, 34

Tests of hypothesis, about,
θ (Bernoulli), 15
θ(exponential), 19, 111
θ_1/θ_2, 38
μ (normal mean), 7, 9
μ_1 and μ_2, 30, 31, 33
σ^2 (normal variance), 11
σ_1^2 and σ_2^2, 34

Total cost, 71

t-test
one-sample, 56
two-sample, 58, 118
two-stage, 10

Two-stage procedure, 4, 10, 17, 25, 26, 32, 33, 46, 49, 64, 76, 94, 112, 114, 117, 121

Statistical Modeling and Decision Science

Gerald J. Lieberman and Ingram Olkin, editors

Samuel Eilon, *The Art of Reckoning: Analysis of Performance Criteria*
Enrique Castillo, *Extreme Value Theory in Engineering*
Joseph L. Gastwirth, *Statistical Reasoning in Law and Public Policy: Volume 1, Statistical Concepts and Issues of Fairness; Volume 2, Tort Law, Evidence, and Health*
Takeaki Kariya and Bimal K. Sinha, *Robustness of Statistical Tests*
Marvin Gruber, *Regression Estimators: A Comparative Study*
Alexander von Eye, editor, *Statistical Methods in Longitudinal Research: Volume I, Principles and Structuring Change; Volume II, Time Series and Categorical Longitudinal Data*
M. M. Desu and D. Raghavarao, *Sample Size Methodology*